AN EXHIBITION HISTORY OF VICTORIAN LEEDS

AN EXHIBITION HISTORY OF VICTORIAN LEEDS

REBECCA WADE

LIVERPOOL UNIVERSITY PRESS

PAUL MELLON CENTRE
for Studies in British Art

First published 2023 by
Liverpool University Press
4 Cambridge Street
Liverpool
L69 7ZU

British Library Cataloguing-in-Publication data
A British Library CIP record is available

ISBN 978-1-80207-854-1 cased

Typeset by Carnegie Book Production, Lancaster
Printed and bound by CPI Group (UK) Ltd, Croydon CR0 4YY

For my friends at Leeds Museums and Galleries,
in memory of John Roles (1957–2020)

Contents

List of figures xi

Acknowledgements xv

Introduction 1

1 The Leeds Public Exhibition, 1839 17

Polytechnic knowledge 19

Paintings, works of art and curiosities 22

Antiquities and natural history 27

Philosophical, chemical and electrical apparatus 29

Machinery, models, materials and manufacture 30

'For the benefit of the Mechanics' Institution' 31

A self-interested philanthropy 34

Conclusion 39

2 The Second Leeds Public Exhibition, 1843 41

Soliciting donations 43

Installation and promotion 46

Previews and reviews 48

The Museum of Antiquities 50

The Picture Gallery 52

The Cabinet 58

The Saloon 65

'Machinery in active operation' 67

Additions and subtractions 68

The second season 71

Conclusion 74

3 The Third Leeds Polytechnic Exhibition, 1845 77

Same difference 80

Diminishing returns, increasing costs 82

The Hall of Sculpture 85

The Picture Gallery 97

Leeds Art Union 100

The Cabinet 102

The Hall of Caves 105

The Saloon 108

For public walks and baths? 110

Conclusion 114

4 Exhibitions and Art Education in Leeds, 1846–56 117

The Exhibition of French Manufactures at the Leeds School of Design, 1846–47 118

Leeds Academy of Arts, 1853–55 122

The 'Travelling Museum' at the Leeds School of Practical Art, 1855–56 128

Conclusion 135

5 The National Exhibition of Works of Art at Leeds, 1868 139

Leeds General Infirmary 140

Looking to London 142

'Planned for the reception of patients rather than of pictures' 145

Painting and portraiture 145

The Museum of Ornamental Art 149

Sculpture and the Royal Academy of Arts 153

The India Museum 155

Conclusion 156

6 The Yorkshire Exhibition of Arts and Manufactures, 1875 159

'Loyalty and Official Robes' 163

The Coloured Cloth Hall 165

The Edinburgh Hall 167

The opening ceremony 170

The Manufactures Department 173

The Machinery Department 174

The Science Department 177

The Sanitary Department 178

The Fine Art Department 180

Farnley Hall 193

Conclusion 194

7 Towards an Art Gallery for Leeds, 1858–88 199

The establishment of a civic collection at Leeds Town Hall 200

The effect of the National Exhibition of Works of Art at Leeds 208

The Yorkshire Fine Art Society 210

Thomas Walter Harding 225

The Fine Art Gallery and Museum Committee 231

The architectural competition 235

Funding, construction and early acquisitions 239

The opening ceremony 247

The public reception 249

Conclusion 251

Epilogue 255

Bibliography 263

Index 273

Figures

1.1 Joseph Rhodes, *Old Music Hall in Albion St Leeds (North Ante Room) in 1839*, 1839, watercolour, 45 × 35.5 cm. Leeds Museums and Galleries, UK/Bridgeman Images 23

3.1 Edward Radclyffe after Benjamin Sly, *Public Exhibition of Frescoes & Sculpture in Westminster Hall*, 1844, steel engraving. Collection of the author 87

3.2 Anonymous, *Grand Staircase and Vestibule of Goldsmiths' Hall*, from *The Illustrated London News*, 13 January 1844, wood engraving. Collection of the author 93

4.1 Anonymous, *Sèvres China from the Museum of Art Manufactures, at Marlborough House*, from *The Illustrated London News*, 18 September 1852, wood engraving. Collection of the author 133

5.1 Anonymous, *Directors and Officials, Exhibition at New Infirmary Leeds 1868*, 1868, photograph. Leeds Museums and Galleries, UK/Bridgeman Images 143

5.2 Anonymous, *Exhibition of Works of Art at Leeds: The Central Hall*, from *The Illustrated London News*, 28 August 1868, wood engraving. Collection of the author 144

5.3 Block cut by Mason Jackson, *The New Infirmary at Leeds, Now Occupied by the National Art-Exhibition, Opened by the Prince of Wales on Tuesday*, from *The Illustrated London News*, 23 May 1868, wood engraving. Collection of the author 146

5.4 Block cut by Mason Jackson, *The Leeds Exhibition: The Gallery of British Living Painters in Oil*, from *The Illustrated London News*, 27 June 1868, wood engraving. Collection of the author 147

5.5 Block cut by Mason Jackson, *National Exhibition of Art at Leeds: The Mediæval Museum*, from *The Illustrated London News*, 11 July 1868, wood engraving. Collection of the author 150

5.6 Block cut by Mason Jackson, *National Exhibition of Works of Art, Leeds: The Museum of Ornamental Art*, from *The Illustrated London News*, 8 August 1868, wood engraving. Collection of the author 151

5.7 Block cut by Mason Jackson, *The Grand Staircase at the Leeds Exhibition of Arts*, from *The Illustrated London News*, 20 June 1868, wood engraving. Collection of the author 154

5.8 Season ticket for the National Exhibition of Works of Art, Leeds, 1868, card. Leeds Museums and Galleries, UK/ Bridgeman Images 157

6.1 Print made by Charles Roberts, *The Duke of Edinburgh Opening the Yorkshire Exhibition of Arts and Manufactures at Leeds*, from *The Illustrated London News*, 22 May 1875, wood engraving. Collection of the author 171

6.2 Anonymous, *Machinery Annexe C, No.1, Yorkshire Exhibition of Arts and Manufactures, Leeds 1875*, 1875, photograph. Leeds Museums and Galleries, UK/ Bridgeman Images 175

6.3 Certificate awarded to William Lupton and Company Limited, 1875. Leeds Museums and Galleries, UK/ Bridgeman Images 176

6.4 Italian School, *Saint Sebastian Tended by the Holy Women*, 17th century, oil on canvas, 168.9 × 108 cm. Leeds Museums and Galleries, UK/Bridgeman Images 190

7.1 William Ritchie and Sons Limited, *Leeds Town Hall*, c. 1913, photographic postcard. Collection of the author 201

7.2 Print made by John Walmsley after Matthew Noble, block cut by Harden Sidney Melville, *Statue of Queen Victoria in the New Towhhall [sic] Leeds*, from *The Illustrated London News*, 11 September 1858, wood engraving. Collection of the author 203

7.3 W.E. Hodgkin, *Leeds New Mechanics' Institution and School of Art*, from *The Illustrated London News*, 21 September 1867, wood engraving. Collection of the author 209

7.4 Anonymous, *The New Municipal Buildings and Free Public Library at Leeds: The New Building*, from *The Graphic*, 26 April 1884, wood engraving. Collection of the author 221

7.5 Anonymous, *The New Municipal Buildings and Free Public Library at Leeds: The Reading Room in the Free Public Library*, from *The Graphic*, 26 April 1884, wood engraving. Collection of the author 232

7.6 William Ritchie and Sons Limited, *Leeds City Art Gallery*, c. 1916, photographic postcard. Collection of the author 243

7.7 F.W., *Municipal Art Gallery and Museum, Leeds*, from *The Illustrated London News*, 27 October 1888, wood engraving. Collection of the author 244

7.8 William Ritchie and Sons Limited, *Queen's Room, Leeds City Art Gallery*, c. 1915, photographic postcard. Collection of the author 245

8.1 F.W., *Bird's-Eye View of the Leeds International Exhibition*, from *The Illustrated London News*, 1890, wood engraving. Collection of the author 257

Acknowledgements

Written during the Covid-19 pandemic, this book is close to home. The support provided by the people and institutions I acknowledge here was not just helpful, but truly sustaining through the most challenging of circumstances. My first thanks must go to the Paul Mellon Centre for Studies in British Art, whose swift and generous response to the crisis made it possible to write the majority of this book during a six-month Research Continuity Fellowship. Alison Welsby and the team at Liverpool University Press and Carnegie Book Production have been wonderful to work with and I very much appreciate their support for this project. I would like to extend my gratitude to the two anonymous readers of the first draft of the manuscript; their advice was invaluable and the book is better for it. I thank my former colleagues at Leeds Museums and Galleries for their assistance in identifying relevant objects from the collections and kindly providing illustrations for this publication. I am grateful to Ann Compton for her encouragement and for pointing me in the direction of useful sources and to Jason Edwards for putting me forward for an Honorary Fellowship with the Department of History of Art at the University of York, which provided access to academic library resources at a crucial time. My family and friends have been steadfast in their support, especially Emii Alrai, Ebony Andrews, Danielle Child, Megan Jones, Claire Mayoh, Kate Nichols, Clare O'Dowd and Mark Westgarth. My deepest thanks are reserved for my wife, Anne-Louise, who gave me not just a person but a whole city to love.

Introduction

At the time of writing Leeds is a year away from staging 'A Year of Culture' in 2023. First conceived as a proposal to be European Capital of Culture in the same year, the United Kingdom's decision to leave the European Union made the city ineligible to apply after the process had already begun. As a great deal had already been invested in the bid, Leeds City Council decided to organise a major international festival regardless. Despite the confident and celebratory rhetoric from the local authority and its partners, there remains an underlying sense that it presents an opportunity for the city to atone for a longstanding cultural deficit. The late head of Leeds Museums and Galleries, John Roles (1957–2020), was fond of saying, 'Leeds doesn't just hide its light under a bushel, it hides the bushel as well.' In contrast to its regional neighbours, particularly Manchester, Liverpool and Birmingham, Leeds has struggled for a cultural status and identity to match its economic success ever since its early industrialisation. Two episodes, exactly a century apart, have come to represent the perceived inferiority of the city's relationship with the visual arts. The first was Leeds's repeated failure to capitalise on the momentum generated by successive temporary exhibitions to build a permanent, purpose-built art gallery until the relatively late date of 1888; and the second, the decision in 1988 to reject the proposed *Brick Man* public sculpture by the artist Antony Gormley, which would have stood ten metres taller than his iconic *Angel of the North* completed for Gateshead a decade later. But delays and missed opportunities, though they may be embedded in the collective memory of the city, do not properly constitute its history.

Beginning in 1839 and ending in 1890, *An Exhibition History of Victorian Leeds* provides the first account of the cultural history of this major British city through its public exhibitions. It situates the city's relationship with the display of fine and decorative arts, industrial culture and the sciences in its regional, national and international contexts and establishes its significance to the history of British art. Too often positioned as minor provincial rehearsals for the 1851 Great Exhibition of the Works of Industry of All Nations, this body of research demonstrates that networks of donation and display were firmly established in Leeds before 1851. These networks continued to develop and extend according to local initiatives and priorities, alongside a keen engagement with the latest metropolitan and international activities. This book investigates the political, economic and social contexts of a distinctive exhibitionary culture alongside its associated individuals and voluntary societies. More broadly, it contributes to the understanding of practices of collecting and exhibiting across social classes, the emergence of museum culture and the negotiation between private and public interventions in cultural politics at both local and national levels.

The history of exhibitions in Victorian Leeds is an urban history. Although Leeds was not granted city status until 1893, by 1851 the town was the fifth largest urban centre in England and the commercial capital of Yorkshire. Its governance, population, institutions, infrastructure and industries were those of a city in all but name, as were its challenges and inequalities. As the historian Asa Briggs established through his influential study of Victorian cities, these urban spaces provided a locus for the formation of distinctive social identities; the organisation of specialised voluntary groups; the publication of newspapers and attendant construction of public opinion and new forms of cultural activity.[1] More recently, urban historians, including Richard Rodger and Robert Colls, have articulated the ways in which cities organise socio-economic relations.[2] They demonstrated that urbanisation often occurred before the provision of appropriate infrastructure. The development and distribution of municipal infrastructure was often determined by the increasing political role of the middle classes, designed to benefit both their industrial and commercial interests in the city and their domestic lives in the growing suburbs.[3] Until the opening of Leeds Art Gallery in 1888, the architectural infrastructure for public exhibitions in Leeds relied on existing buildings designed for other purposes: a music hall, a warehouse, a hospital, a cloth hall. In

[1] Asa Briggs, *Victorian Cities* (London: Odhams, 1963), p. 24.

[2] Richard Rodger and Robert Colls, 'Civil Society and British Cities', in *Cities of Ideas: Civil Society and Urban Governance in Britain 1800–2000*, ed. by Richard Rodger and Robert Colls (Aldershot: Ashgate, 2004), pp. 1–20 (pp. 1–2).

[3] Rodger and Colls, 'Civil Society and British Cities', p. 3.

the first half of the period the social and intellectual infrastructure was appropriated too, from existing voluntary societies, fundraising bodies and municipal structures. The concurrent development of transport infrastructure was particularly significant for public exhibitions in Victorian Leeds. New railways connected regional and national audiences to the town and provided the means to transport whole collections of objects across the country with greater speed and safety than had previously been possible. The Leeds–Selby line opened in 1834 and was extended to Hull six years later. The North Midland Railway opened the Derby–Leeds line in 1840, which connected Leeds to London by rail for the first time. A year later, the Manchester–Leeds line opened, and in 1846 Leeds was connected with neighbouring Bradford.[4] In the second half of the century new stations were built closer to the centre of the town and the lines established by competing companies were increasingly integrated. Alongside improvements to existing routes by river, canal and road, the railways stimulated demographic change and economic growth, making inter-urban travel more accessible. The organisers of public exhibitions in Leeds were cognisant of the potential to bring large numbers of visitors from beyond the town and quickly established special trains and affordable fares to encourage excursions from the wider region and beyond.

At the beginning of the period in question the town had upwards of one hundred woollen mills and thirty flax spinners.[5] By mid-century the composition of industrial activity in Leeds had already diversified, with engineering having emerged as a substantial ancillary to the textiles industry. Unlike Manchester in neighbouring Lancashire, Leeds was not as dependent on the import of cotton as the chief raw material processed in its factories, and as such, its wealth was not primarily built on the labour of enslaved people and the triangular trade. As a result the wealth generated in Leeds—although still reliant on exploitation and embedded in colonial structures—was generally of a different order of magnitude to the massive fortunes made in Manchester or the ports of Liverpool, Bristol, Glasgow and London. This had a substantial effect on the complexion of the cultural and moral life of the town, which tended not to be driven by the philanthropy of named individuals to the same extent.

4 R.W. Unwin, 'Leeds Becomes a Transport Centre', in *A History of Modern Leeds*, ed. by Derek Fraser (Manchester: Manchester University Press, 1980), pp. 113–41 (pp. 132–33).

5 Asa Briggs, 'Leeds: A Study in Civic Pride', in *Victorian Cities*, pp. 139–83 (pp. 140–41).

There is a paucity in the secondary literature on temporary exhibitions outside London, and the research that does exist tends to have been published as short journal articles. The literature on nineteenth-century temporary exhibitions has been dominated by the Great Exhibition of 1851, and as a result, regional exhibitions that took place in the second half of the century are almost invariably framed in relation to it. Recent scholarship on the emergence of municipal galleries and museums in Britain in the last quarter of the nineteenth century has positioned these exhibitions merely as the means by which permanent institutions were founded. While the various organisers of temporary exhibitions in Leeds often expressed the hope that their successes would result in a permanent public art gallery, the establishment of such an institution was rarely stated as a primary aim. This book shows that the formation and maintenance of middle-class identities, the extension of the activities of local voluntary societies, civic rivalry and commercial advantage were much stronger motivating factors to stage temporary public exhibitions.

The Reform Acts resulted in the creation of two parliamentary seats for the Borough of Leeds, extended to three under the Representation of the People Act in 1867 and five after the Third Reform Act of 1884.[6] The cumulative effect was to reconceptualise Leeds as 'a legally defined socio-geographic unit of power'.[7] The wider implication was broader political representation, although it remained exclusive and dependent on economic status and gender. For the historian Dror Wahrman, the Reform Acts cast the middle classes as the primary agents of social and political life, and their centrality to the public life of Leeds will be evident throughout this book.[8] The town was divided into wards, each with its own particular history and demographic. Almost every individual and institution associated with public exhibitions in Leeds in the first half of the nineteenth century was situated in the ward of Mill Hill, with its concentration of bankers, lawyers, politicians and medical men.[9] The area was bounded to the west by Park Square, Albion Street to the east and Park Row to the north. By the period in question Mill Hill had been absorbed into the expanding town centre and its position made it vulnerable to the encroachment of industry. Its proximity to

[6] 'Leeds', *Graphic*, 18 July 1885, pp. 73–79 (p. 75).

[7] R.J. Morris, *Class, Sect and Party: The Making of the British Middle Class, Leeds 1820-1850* (Manchester: Manchester University Press, 1990), p. 127.

[8] Dror Wahrman, *Imagining the Middle Class: The Political Representation of Class in Britain, c. 1780–1840* (Cambridge: Cambridge University Press, 1995), pp. 17–18.

[9] Morris, *Class, Sect and Party*, p. 14.

the factories along the River Aire to the south meant that its wealthiest inhabitants chose to migrate north, leaving the Georgian Park Estate and its environs to the middle classes.[10]

The concepts of social class, voluntary action and the relationships between Leeds and its local, national and international networks are central to this study. The emergence and diffusion of middle-class culture and identities were particularly important to exhibition making during this period, but it cannot be considered a stable category. The varied composition of the British middle classes has even been credited with their dominance in the nineteenth century.[11] The emergence of the middle classes was connected to the growth of industrial capitalism and it was the middle classes who drove the development of industrialisation in urban centres.[12] However, industrialisation, uneven in its progress and effects, did not result in a clean distinction between a class of employers and wage labourers. In early Victorian Leeds, for instance, a relatively small number of manufacturers employed a mass of wage labourers.[13] Their numbers may have been limited, but the influence they exerted over public life was disproportionate and exhibitions formed a means of extending their paternalistic position towards their workforce.[14] Shared culture bound the disparate middle classes together: 'At the art exhibition, the concert hall and the social club, men—and, in some cases, women—could engage in activities that were deemed to transcend the divisions of sect and party.'[15] The project of voluntary societies such as the Leeds Mechanics' Institution and the Leeds Philosophical and Literary Society was to foster social unity through collective culture. The Leeds Philosophical Society was established in 1819 and had by 1821 built the Philosophical Hall to house a museum, lecture theatre, library, laboratory and meeting rooms on Park Row. The Society facilitated temporary public

[10] Morris, *Class, Sect and Party*, p. 39.

[11] Alan Kidd and David Nicholls, eds, *Gender, Civic Culture and Consumerism: Middle-Class Identity in Britain, 1800–1940* (Manchester: Manchester University Press, 1999), p. xxv.

[12] Simon Gunn, 'The Public Sphere, Modernity and Consumption: New Perspectives on the History of the English Middle Class', in *Gender, Civic Culture and Consumerism: Middle-Class Identity in Britain, 1800–1940*, ed. by Alan Kidd and David Nicholls (Manchester: Manchester University Press, 1999), pp. 12–29 (p. 12).

[13] Morris, *Class, Sect and Party*, p. 22.

[14] Caroline Arscott, '"Without distinction of party": The Polytechnic Exhibitions in Leeds 1839–45', in *The Culture of Capital: Art, Power and the Nineteenth-Century Middle Class*, ed. by Janet Wolff and John Seed (Manchester: Manchester University Press, 1988), pp. 135–58 (p. 151).

[15] Gunn, 'The Public Sphere, Modernity and Consumption', p. 24.

exhibitions in Leeds in a number of ways, most often through the loan of specimens from its growing collections.[16] These voluntary societies were flexible enough to accommodate a variety of positions and levels of activity, which enabled them to act together in support of their shared social, political and economic interests.[17] This also meant that stratification persisted through the organisational structures of voluntary societies. At the top of the membership hierarchy were a small number of proprietors, who in the case of the Leeds Philosophical and Literary Society secured their position with the payment of £100 for a share in the construction of the Philosophical Hall.[18] Although patronage from the gentry was refused, the substantial financial commitment ensured that only the upper middle classes could attain this status.[19] The hierarchy could be a cause of tension as a small set of people tended to govern over an extended period of time, as will become clear over the following chapters as we continue to encounter the same influential individuals and families across the whole period in question.[20] The complexities of these internal relations were matched, if not exceeded, by the complexities of relations with other societies and, in turn, their interactions with the working classes.

The Leeds Mechanics' Institution mediated between the middle-class composition of the Leeds Philosophical and Literary Society and the working classes of the town. However, of the twenty members of the first committee of the Mechanics' Institution, thirteen members were also on the committee of the Philosophical and Literary Society.[21] The primary function of the Mechanics' Institution was to provide technical education to a relatively new social group of skilled operatives whose labour was shaped by mechanised production.[22] In its early years the Leeds Mechanics' Institution operated from rented rooms behind Park Row, which included a large room for lectures and 'a tolerably good room for a library ... in an exceedingly confined and remote situation'.[23] The reports from the early 1830s show that it was a priority to maintain

[16] Edwin Kitson Clark, *The History of 100 Years of Life of the Leeds Philosophical and Literary Society* (Leeds: Jowett and Sowry, 1924), p. 52.

[17] Morris, *Class, Sect and Party*, p. 167.

[18] Morris, *Class, Sect and Party*, p. 236.

[19] Morris, *Class, Sect and Party*, p. 238.

[20] Samuel Alberti, 'Natural History and the Philosophical Societies of Late Victorian Yorkshire', *Archives of Natural History*, 30:2 (2003), 342–58 (p. 343).

[21] Morris, *Class, Sect and Party*, p. 230.

[22] *The Leeds Institute of Science, Art and Literature: Historical Sketch, 1824–1900* (Leeds: Goodall and Suddick, 1901), p. 5.

[23] Edward Parsons, *The Civil, Ecclesiastical, Literary, Commercial, and*

premises on or near Park Row in Mill Hill, close to the Philosophical Hall and the cultural and commercial centre of the town. Proximity to the headquarters of the Philosophical and Literary Society was mutually beneficial: the Mechanics' Institution benefited from the patronage of its more established, wealthier neighbour and the Philosophical and Literary Society from the opportunity to extend its influence to a working-class public.[24] Temporary exhibitions were a site of visible cooperation between the two groups.[25]

In addition to official catalogues and descriptive guides, the major source of primary material on the culture of temporary exhibitions in Leeds is the local periodical press. In her critical analysis of Victorian journalism, Lyn Pykett cautioned against interpreting these sources as unmediated historical evidence. Newspapers and journals did not simply represent the culture on which they reported; they were an active part of it. For this reason, Pykett advocated a wider exploration of the context in which they were published.[26] The *Leeds Mercury* provided the most detailed and sustained coverage of the exhibitions explored in this book, so it is important to understand the politics of its proprietors. Edward Baines (1774–1848) purchased the *Leeds Mercury* in 1801. In his history of the Leeds Philosophical and Literary Society Edwin Kitson Clark wrote that Baines had 'done more for the cause of Reform in the County of York than any other man' and articulated his beliefs in the following terms: 'an advocate of public economy, an enemy of monopoly, an opponent of civil disqualification of dissenters, and a supporter of the emancipation of slaves'.[27] During the 1820s his son Edward Baines Junior (1800–90) extended his editorial influence over the *Leeds Mercury*, which had been used a means of campaigning for political representation under his father's tenure.[28] Once Baines Senior was elected a Member of Parliament for Leeds in 1834, his son took complete responsibility for the newspaper and continued to use his position to advance a similar set of social, political and economic causes as his father had pursued. Against state intervention—especially

Miscellaneous History of Leeds, Bradford, Wakefield, Dewsbury, Otley, and the District within Ten Miles of Leeds (Leeds: F. Hobson, 1834), p. 104.

24 Alberti, 'Natural History and the Philosophical Societies', p. 351.

25 Morris, *Class, Sect and Party*, p. 233.

26 Lyn Pykett, 'Reading the Periodical Press: Text and Context', in *Investigating Victorian Journalism*, ed. by Laurel Brake, Aled Jones and Lionel Madden (Basingstoke: Macmillan, 1990), pp. 3–18 (p. 7).

27 Kitson Clark, *100 Years of Life of the Leeds Philosophical and Literary Society*, p. 12.

28 Wahrman, *Imagining the Middle Class*, p. 400.

in matters of education, religion and industrial regulation—Edward Baines Junior strongly promoted the public exhibitions of Leeds as an extension of his belief in 'voluntaryism' and its opportunities for self-improvement. He was also embedded in the organisation of the exhibitions associated with the Leeds Philosophical and Literary Society and Leeds Mechanics' Institution, as a founder member of the former and longstanding president of the Yorkshire Union of Mechanics' Institutes.

The relationship between industrialisation, education and public exhibitions has been the subject of new scholarship in recent years, which often traces the influence of the pre-eminent art critic John Ruskin (1819–1900) on a generation of regional reformers who sought to instrumentalise the visual arts to ameliorate poor living and working conditions in urban centres.[29] Ruskin was not a dominant voice in the cultural life of Victorian Leeds, which was instead informed by the concepts of 'voluntaryism' and 'self-help' articulated locally by Edward Baines Junior and Samuel Smiles (1812–1904) respectively. The 'Bainesocracy' was a pejorative term used to describe the collective dominance of the Baines family over political discourse in the town.[30] Samuel Smiles moved to Leeds in 1838 to become the editor of the *Leeds Times*, a radical newspaper in competition with Baines's *Leeds Mercury*.[31] Rather than his later and more famous work, *Self-Help: With Illustrations of Character and Conduct* of 1859, it is more appropriate in this regional context to consider its precursors, which included a lecture delivered to the Bradford United Reform Club at the Bradford Mechanics' Institution on 14 February 1842, titled *The Diffusion of Political Knowledge among the Working Classes*.[32] Smiles's time in neighbouring Leeds was particularly significant to the formation of his philosophy of autodidacticism.[33]

Baines and Smiles occupied different political positions, which shifted over time and in response to particular campaigns. Their beliefs and values intersected on a number of important points, which

[29] See Amy Woodson-Boulton, *Transformative Beauty: Art Museums in Industrial Britain* (Palo Alto, CA: Stanford University Press, 2012).

[30] J.F.C. Harrison, 'Chartism in Leeds', in *Chartist Studies*, ed. by Asa Briggs (London: Macmillan, 1959), pp. 65–98 (pp. 66–67).

[31] Smiles was the editor of the *Leeds Times* from January 1849. See Harrison, 'Chartism in Leeds', p. 83.

[32] Samuel Smiles, *The Diffusion of Political Knowledge among the Working Classes. An Address Delivered before the Members of the Bradford United Reform Club, February 14, 1842* (Leeds: Buckton, 1842).

[33] Kenneth Fielden, 'Samuel Smiles and Self-Help', *Victorian Studies*, 12:2 (1968), 155–76 (p. 168).

included their rejection of the principle of universal suffrage.[34] The ideology of the voluntary society and 'the Utopian dream of a wise and cultivated workforce' provided common ground and a moral and economic rationale for public exhibitions in the town.[35] This social project has been characterised as inherently paternalistic in its attempts to impose the cultural values of the urban middle classes on the working classes.[36] Smiles regretted the middle-class governance and membership of mechanics' institutions generally and was particularly troubled by the merger of Leeds Mechanics' Institution with the Leeds Literary Institution in 1842.[37]

The tacit function of voluntary societies such as the reconstituted Leeds Mechanics' Institution and Literary Society was the construction and maintenance of middle-class identity, distinct from the working classes below and gentry above.[38] As part of this collective project, education was afforded an exalted status, offering transformative possibilities for the individual and society. Smiles described his desire 'to elevate the moral and intellectual condition of the mass, as to fit every individual member of society for the performance of his duties as a social and intelligent being'.[39] In one of several papers read before the Leeds Philosophical and Literary Society, Baines offered the following assessment of the importance of education:

> It has been truly said, that man is the creature of education. Although no artificial means can add to the number either of his intellectual or corporeal faculties, yet both allow wonderfully different degrees of improvement in their nature & skill in their employment. The mind, in particular, may be so far contracted by ignorance, so far enlarged & exalted by knowledge, that experience

34 Harrison, 'Chartism in Leeds', p. 84.

35 Adrian Rifkin, 'Success Disavowed: The Schools of Design in Mid-Nineteenth Century Britain. (An Allegory)', *Journal of Design History*, 1:2 (1988), 89–102 (p. 89).

36 R.J. Morris, 'Voluntary Societies and British Urban Elites 1780–1850', in *The Eighteenth-Century Town*, ed. by Peter Borsay (London: Longman, 1990), pp. 338–66 (p. 363).

37 Smiles, *The Diffusion of Political Knowledge*, p. 17. The Leeds Literary Institution was founded in 1834 and operated at the Commercial Buildings until 1842, when it joined the Mechanics' Institution at 12 South Parade. See *The Leeds Institute of Science, Art and Literature: An Historical Sketch of One Hundred Years' Work, 1824–1923* (Leeds: Jowett and Sowry, 1923), p. 7.

38 Alberti, 'Natural History and the Philosophical Societies', p. 352.

39 Smiles, *The Diffusion of Political Knowledge*, p. 8.

> only would enable us to recognize, in two individuals who display its greatest diversity, creatures of the same species & of equal native powers.[40]

In addition to a base level of innate intelligence, Baines proposed further conditions necessary for individual and social advancement, in part based on observations made about the perceived superiority of ancient Athenian art, science and judicial structures. Economic prosperity was one condition that Baines could claim for Leeds when he suggested that 'literature & science cannot flourish in any country, till a considerable proportion of the population of the inhabitants have acquired affluence, & can enjoy leisure'.[41] This view relied on his assertion that wages in the manufacturing districts were higher than in other areas of the country, as the woollen industry had not been subject to the same levels of mechanisation as others, including the production of cotton associated with Yorkshire's longstanding rivals in Lancashire.[42] Baines estimated the average weekly wage in the woollen industry to be 12s. 6d., although the age, gender and particular role of the worker altered the figure considerably.[43]

Connected to the notion that economic prosperity could foster cultural and intellectual progress was the belief in competition to drive advancement at every level from the individual to the nation state. Baines attributed the success of the ancient Athenians to the ways in which competition permeated daily life in the struggle to attain and maintain citizenship. He particularly emphasised the beneficial effect on the arts: 'competition furnishes the greatest stimulus to excellence of almost every kind: to this, more than to any thing else, is the world indebted for all the works of art and the productions of talent'.[44] In

40 Edward Baines, 'On the Rise of Art, Science and Liberation among the Athenians', read before the Leeds Philosophical and Literary Society on 12 April, 1822, p. 1. Papers of Edward Baines the Elder and Edward Baines the Younger, West Yorkshire Archive Service, Leeds, WYL383/83/49.

41 Baines, 'On the Rise of Art, Science and Liberation among the Athenians', p. 15.

42 Edward Baines, *On the Social, Educational and Religious State of the Manufacturing Districts*, 3rd edn (London: Simpkin, Marshall & Co. and T. Ward & Co., 1843), p. 55.

43 Edward Baines, 'On the Woollen Manufacture of England, with Special Reference to the Leeds Clothing District', *Journal of the Statistical Society of London* (1859), 1–34 (p. 28).

44 Edward Baines, 'On the Elgin Marbles and the Causes of the Excellence of Grecian Sculpture'. Unpaginated. Personal Papers of Edward Baines the Elder and Edward Baines the Younger.

contrast, Smiles advocated collaboration above competition: 'the first great test of a civilized being is, in that he is able to co-operate or combine with others, for the purpose of obtaining some object which is desirable alike to all'.[45] For Smiles, collaboration between the social classes was critical to the success of civil society.[46]

In addition to economic prosperity and competition, Baines specified the optimum political conditions necessary for progress: 'nothing, so much as Liberty, promotes the intellectual advancement, the mental vigour, the virtue, the bravery, the wealth, & the power of nations'.[47] The meaning of liberty in this context encompassed a range of positions and interests. Baines used this term in relation to individual, religious or regional self-determination, to resist what was perceived as interference by the state. Baines saw constant threats to liberty in the form of national initiatives and standardised procedures: 'in the matter of education, as of government, liberty is subject to irregularities, which are an eye-sore to the martinets of despotism, as well as to men who have imbibed from Continental examples the *doctrinaire* love of uniformity and centralization'.[48] Smiles concurred with this position:

> a Government, no matter how good, can do little or nothing towards making the idle and slovenly industrious, the improvident careful, the drunken sober, the lewd virtuous. No! These are reforms which must spring form the people themselves ... it is sheer folly to look to government for that which we have the means of doing for ourselves. Far more is always to be done by awakening and exciting the spirit of the people, than by the means of the tardy and extorted acts of the legislature ... if we leave it to a central government to decide what shall be the education of the people, we virtually give that government the power of directing public opinion, of stereotyping the national intellect, and of maintaining in the minds of the masses, certain determinate forms of thought, which may contain in them the germs of even the vilest forms of political and religious slavery.[49]

[45] Smiles, *The Diffusion of Political Knowledge*, pp. 11–12.
[46] Smiles, *The Diffusion of Political Knowledge*, pp. 15–16.
[47] Baines, 'On the Rise of Art, Science and Liberation among the Athenians', p. 19.
[48] Edward Baines, *An Alarm to the Nation on the Unjust, Unconstitutional and Dangerous Measure of State Education Proposed by the Government* (London: Ward & Co. and Simpkin, Marshall & Co., 1847), p. 3.
[49] Smiles, *The Diffusion of Political Knowledge*, p. 9.

Baines advocated self-governance or 'voluntaryism', allied with civic responsibility, correct moral behaviour and rational judgement. Baines made a case for this system with the following statement:

> The advantages of self-government are so generally admitted, that all the modern reforms have been designed to make it more absolute. Not that self-government, either in a nation or a municipality, is a faultless system, but that on the whole it is incomparably the best system, because under it lives the ennobling spirit of Liberty. It has taught the people to look less to their governors, and more to themselves. It has given them the invaluable habit of self-reliance.[50]

In practice the relation between the centre and the periphery operated on a mixed economy of private and public, voluntary initiatives and prescriptive order. The principle of self-improvement was deeply embedded in the ideals of the voluntary society, although the middle classes were more interested in protecting their position than inviting the working classes to join them in their social, political and economic advantages. The voluntary society provided a means for the owners of the means of production to maintain a distanced proximity to their wage labourers in order to retain and extend their influence over them.[51] Smiles identified an awareness and dissatisfaction with this situation among the working classes:

> there is something about the management of these institutions which is by no means agreeable to the operative classes generally. They are supported mainly by patronage, and conducted by patronage also. Now, the intelligent portion of the working classes at the present day, hate patronage of any kind. They are in love with self-government and self-governing institutions.[52]

The arguments for voluntaryism in Leeds were characterised by the public pronouncements of Edward Baines Junior and Samuel Smiles. Although they both presented a forceful defence of self-governance, the extension of the state alongside local eagerness for its resources resulted in an uneasy union. In the second half of the nineteenth century, public exhibitions in Leeds increasingly drew on collections

[50] Baines, *An Alarm to the Nation*, p. 3.

[51] Morris, 'Voluntary Societies and British Urban Elites', p. 356.

[52] Smiles, *The Diffusion of Political Knowledge*, p. 18.

and expertise located outside Yorkshire. In addition to having actively sought and accepted government grants towards the establishment and maintenance of the Leeds School of Design in the 1840s, state aid was supplied to exhibitions in Leeds in the form of loans from the South Kensington Museum and its antecedents, and support from the associated government body, the Department of Science and Art. These metropolitan institutions were invested in the dissemination of their design reform agenda through regional exhibitions and the formation of permanent art galleries and museums on their own model. The display of collections from beyond the borough also provided a means for exhibition organisers to present objects that had not yet been seen in Leeds. The exhibitions of the previous decades tended to draw on some of the same local sources, which risked repetition and an associated loss of support from visitors.

Beginning in 1839 with the first Leeds Public Exhibition at the Albion Street Music Hall, the book is arranged chronologically by exhibition until the final chapter, which traces the development of a permanent, purpose-built art gallery for Leeds over the course of three decades. The Leeds Public Exhibition of 1839 was organised by a committee principally composed of members of the Leeds Philosophical and Literary Society and the Leeds Mechanics' Institution. These voluntary societies are discussed in relation to the formation of middle-class identities, the production of knowledge and the maintenance of social order. The ways in which classificatory systems were employed and reinforced in the Leeds Public Exhibition are examined and compared with other examples of polytechnic institutions. The site of the exhibition is considered in the context of the development of public culture, civic values and behaviours. In addition, the didactic intention of the exhibition and the mechanisms through which knowledge was transmitted and stratified are analysed alongside wider debates concerning the improvement of manufactured commodities.

Chapter 2 looks at the Second Leeds Public Exhibition, held again at the Albion Street Music Hall, four years after the first iteration in 1843. It provides an account of the development of the exhibition and the importance of inter-city rivalry, the committee having been spurred into action by recent exhibitions in Hull, Liverpool and Manchester. It investigates the stated intentions of the organisers and the process of soliciting objects for display. The chapter goes on to describe the contents of the exhibition in the order they were intended to be visited, alongside an analysis of the commentary on the exhibition in the periodical press and the reception of contemporary British art. This chapter argues that the Second Leeds Public Exhibition both conformed to and departed from the model set by the first exhibition, with a number of the same organisers and lenders involved once again. Important differences, such as the physical separation of the steam engines and working models from the rest of the display, increased the visibility of performed labour but began to detach the processes of industry from the fine arts and sciences. It is further argued that the 1843 exhibition cannot be considered a fixed or static event: objects flowed in and out of the galleries throughout the duration of the exhibition, which reopened for a second season with a different purpose and character after its initial closure to the public.

The third chapter takes the final Leeds Polytechnic Exhibition of 1845 as its subject. Unlike previous iterations, it was principally organised by four trustees: the Mayor of Leeds, the Vicar of Leeds and the two Members of Parliament for the borough. In addition to soliciting loans from a broadly similar base of lenders, this exhibition drew substantially from the pool of works generated through the Fine Arts Commission's task of ornamenting the new Houses of Parliament with paintings and sculpture. Chapter 3 argues that the Third Leeds Polytechnic Exhibition—rather than being a rehash of the previous two exhibitions and the victim of diminishing returns—represented a distinct development, especially in the display and interpretation of contemporary British sculpture, the publication of a dedicated journal and formation of an art union. Besides its content, the exhibition also had a fundamentally different purpose and organisational structure: to raise funds for public walks and baths rather than to benefit the Leeds Mechanics' Institution, thereby directly engaging with the urgency of contemporary sanitary reform.

Chapter 4 combines three exhibitions staged in Leeds during the decade between 1846 and 1856. Beginning with the Exhibition of French Manufactures at the new Leeds School of Design, it goes on to narrate the establishment and dissolution of the Leeds Academy of Arts and its exhibition programme, before considering the display of the 'Travelling Museum' at the Leeds School of Practical Art between 1855 and 1856, composed of the recently formed Circulating Collection from the Museum of Ornamental Art at Marlborough House, which included examples of Italian majolica, German and Flemish stoneware, Chinese porcelain, Russian engraving, French furniture, Turkish and Indian silks and Spanish and Irish lace. It is argued that this period was especially concerned with the relationship between centralised provision of art and design education and Leeds's specific social, political, economic and cultural context.

Chapter 5 provides an account of the National Exhibition of Works of Art, held between May and October 1868 at Leeds Infirmary before it was officially opened as a functional hospital in May 1869. Designed in the Gothic Revival style by George Gilbert Scott, the empty ecclesiastical spaces were appropriated for the display of modern and old master paintings, contemporary British sculpture, a portrait gallery of 'Yorkshire Worthies', a Museum of Ornamental Art composed of objects from the South Kensington Museum and an 'India Museum' with loans from the Museum of the East India Company. This chapter argues that the exhibition deserves to be recognised as one of the most significant of the period, given its scale, content, arrangement and the substantial discourse it generated. It connected Leeds to national and international networks of museum professionals, collectors and art historians while making use of the political, economic and social capital of its own second- and third-generation industrialists. The exhibition operated as a crucial staging post for some of the most significant British painting, sculpture and decorative arts, between the private collections from which they were mostly drawn and the regional, national and international public collections many of these objects would later enter.

Chapter 6 considers the Yorkshire Exhibition of Arts and Manufactures. In May 1874 the committee of the Leeds Mechanics' Institution and Literary Society decided to hold a major exhibition of art and industry to liquidate the remaining debt of £9,500 on the Leeds Mechanics' Institute building. Led by the Mayor of Leeds, Henry Rowland Marsden, the exhibition was to have two departments: Arts and Manufactures and Fine Arts. The Fine Arts department included oil paintings, watercolours and drawings; sculptures, models, die sinking and intaglios; etchings and engravings; antiquities and photography. Opened on 6 May 1875 by the Duke of Edinburgh under the patronage of Queen Victoria, the exhibition was staged at the Coloured Cloth Hall. Spanning a site of 1.5 acres, it was the most ambitious in scale and extent yet seen in Leeds and succeeded in its aim to clear the debt on the Institute. This chapter argues that the Leeds Mechanics' Institution returned to—and extended—the model of the polytechnic exhibitions of the late 1830s and 1840s in an attempt to avoid a repeat of the financial failure of the National Exhibition of Works of Art in 1868. It attributes its success to its having secured visible royal patronage; articulated a clear aim to support the Mechanics' Institution; obtained the support of local social, political and economic elites; and constructed a lineage back to the polytechnic exhibitions of the previous generation. Lastly, the chapter argues that criticism of the exhibition in the metropolitan periodical press was based on its refusal to conform to developing ideas about 'correct' classification and arrangement and its appropriation of an 'ugly' Georgian commercial space.

The final chapter provides an account of the establishment of a permanent civic art collection and purpose-built gallery in Leeds, from the first attempt associated with the new Town Hall in 1858. It follows the activities of the Yorkshire Fine Art Society in the 1870s and 1880s, which leased two floors in the new Athenæum Building to stage nine exhibitions before it was wound up in 1883. The impetus was taken up by Councillor Thomas Walter Harding (1843–1927), a second-generation industrialist, who in 1886 persuaded the Town Council to assign part of the funds raised by public subscription to celebrate the Golden Jubilee of Queen Victoria in 1887 to build the gallery as an extension to the existing Municipal Buildings. Chapter 7 argues that although the establishment of Leeds Art Gallery represented the realisation of long-held civic ambitions, it also marked the end of a distinctive period of temporary exhibition making that appropriated commercial, industrial, healthcare and military spaces to produce and circulate polytechnic knowledge.

1

The Leeds Public Exhibition, 1839

The Leeds Public Exhibition was the first major exhibition to be staged in the town in the new Victorian era. Over the course of its three month duration 183,913 people visited, when the population of Leeds was recorded as 172,720 at the point of the 1841 Census.[1] The exhibition took place between 9 July and 5 October 1839 at the Albion Street Music Hall in the centre of the town.[2] It merits investigation both as a discrete event and as a product of a wider set of social, political and economic conditions, which allow for the exploration of the intersections between art and industry, taste and morality, commerce and culture, civic identity and social space. The existing scholarship on the Leeds Public Exhibition takes the form of brief mentions in journal articles and fleeting appearances in other publications. Articles by R.J. Morris and Toshio Kusamitsu reference the Leeds Public Exhibition alongside comparable exhibitions in Manchester, Birmingham, Newcastle and Sheffield. These exhibitions are positioned as somewhat generalised regional precedents of the Great Exhibition of 1851.[3] The construction

1 Population as recorded in 'Leeds', *Graphic*, 18 July 1885, pp. 73–79.

2 The exhibition of 1839 is also variously referred to in the literature as *The Exhibition of Paintings, Curiosities, Models, Apparatus, and Specimens of Nature and Art* and *The Leeds Public Exhibition of Works of Art, Science, Natural History, and Manufacturing Skill*, alongside other inventive and lengthy variations on the theme. The committee of the exhibition and the *Leeds Mercury* refer to the event as *The Leeds Public Exhibition*. The term Polytechnic was attached to the subsequent exhibitions of 1843 and 1845 and was retrospectively applied by Caroline Arscott in '"Without distinction of party": The Polytechnic Exhibitions in Leeds 1839–45', in *The Culture of Capital: Art, Power and the Nineteenth-Century Middle Class*, ed. by Janet Wolff and John Seed (Manchester: Manchester University Press, 1988), pp. 135–58.

3 R.J. Morris, 'Leeds and the Crystal Palace: A Provincial–Metropolitan Link Bringing Education to Industrial Society', *Victorian Studies*, 13

of this lineage, although it recognises the significance of cultural activities that took place outside the capital, has to some extent denied the Leeds Public Exhibition its specificity. The art historian Caroline Arscott has offered the most sustained critical analysis of the 1839, 1843 and 1845 Leeds Public Exhibitions, set in their social, political, cultural and economic contexts. As with James Lomax's account of the 1843 exhibition, however, there has been a tendency in the literature to shift between conflating the three exhibitions and regarding them as discrete events.[4] Although there are comparisons to be drawn, it is this latter approach that will enable a more comprehensive account of all three exhibitions across these opening chapters.

The archive that stands in for the Leeds Public Exhibition is the catalogue of its contents, a descriptive guide to its objects and coverage in the local press. The price of the descriptive guide was one shilling, against the much lower cost of admission at sixpence, which limited what could be known by whom. The guide was partially serialised in the *Leeds Mercury* during the exhibition, although this periodical was generally written by and for the middle classes. When the descriptive guide was published two months after the opening, it was heavily advertised in the *Leeds Mercury* as both an educational tool and an object of posterity.[5] The guide was positioned as an aid to those who sought to derive 'profit as well as intelligent pleasure' from the objects, to provide meaning and a record of what had been displayed for posterity.[6] The *Leeds Mercury* encouraged its readers to purchase the guide for children as a educational tool 'beyond mere recreation ... to obtain real instruction' ahead of visiting the exhibition in person.[7]

The connection between the Leeds Public Exhibition and the *Leeds Mercury* was not superficial. Its significance is located in the shared interests held by the publication, the committee, the lenders of objects and the beneficiaries of the profits from the exhibition. Most deeply embedded across all of these interests was Edward Baines Junior (1800–90), who was at once the editor of the *Leeds Mercury*, the

(1970), 282–300; Toshio Kusamitsu, 'Great Exhibitions before 1851', *History Workshop*, 9 (1980), 70–89.

4 Arscott, 'Without distinction of party', pp. 135–58; James Lomax, 'Buying Antiques in Early Victorian Leeds: The 1843 Exhibition', *Furniture History*, 33 (1997), 275–85.

5 'Leeds Public Exhibition', *Leeds Mercury*, 7 September 1839, p. 4.

6 William West and Edward Baines, *A Description of some of the Principal Paintings, Machinery, Models, Apparatus and other Curiosities at the Leeds Public Exhibition* (Leeds: Edward Baines and Sons, 1839), p. 1.

7 'Leeds Public Exhibition', *Leeds Mercury*, 21 September 1839, p. 4.

vice-president of the Leeds Mechanics' Institution, a donor of objects and a co-author of the descriptive guide to the exhibition. Although Baines was to some extent exceptional in the extent of his public activity in this context, this interpenetration of interests was present throughout the organisational structure of the exhibition.

Polytechnic knowledge

Polytechnic exhibitions emerged as a distinct cultural form in the second quarter of the nineteenth century. In addition to regional precursors, institutions established in London in the 1830s informed the Leeds Public Exhibition. The National Gallery of Practical Science—also known as the Adelaide Gallery—was established in 1831 and the Polytechnic Institution in 1838.[8] There are compelling similarities between the objects displayed, the ways in which they were interpreted and their arrangement in space. A reference was made in the coverage of the exhibition in the *Leeds Mercury* relating to one of several popular demonstrations: 'in London, we believe the mice in the Adelaide Gallery appear to suffer considerably; but if this be so, it must be from a defect in the Diving Bell or the manner of using it'.[9] This live experiment used a form of early diving equipment to create a pocket of breathable air underwater. Scheduled according to a timetable of events, it involved the safe submersion of small animals in a large model canal in the Saloon of the Music Hall, which was also a central feature of the Adelaide Gallery in London. It was circular in form, with a diameter of 5.5 metres and a depth of 46 centimetres. Miniature steam-powered vessels and goldfish animated the surface of the water, overlooked by a model of Eddystone Lighthouse lit with gas. In the centre of the canal was a fountain that sprayed water several feet into the air before it flowed back into the canal across a waterwheel that generated enough electricity to transmit 'Galvanic Shocks' to brave visitors.[10]

That mechanical, electrical and industrial technologies were presented alongside works of fine and applied art, alongside collections of seemingly indiscriminate ephemera and curiosities, has led some commentators to suggest that this type of exhibition lacked a coherent rationale. The historian Toshio Kusamitsu argued that, 'by comparison with later

[8] Richard Altick, 'Technology for the Million', in *The Shows of London* (Cambridge, MA: Harvard University Press, 1978), pp. 375–89.

[9] 'Leeds Public Exhibition', *Leeds Mercury*, 3 August 1839, p. 5.

[10] *Catalogue of the Exhibition of Paintings, Curiosities, Models, Apparatus, and Specimens of Nature and Art* (Leeds: Edward Baines and Sons, 1839), p. 26.

industrial exhibitions, the mechanic's exhibition had an undifferentiated and heterogeneous character'.[11] In his history of the emergence of public art galleries and museums in Britain, Giles Waterfield was equally dismissive of the arrangement of these early exhibitions: 'This style of event anticipated the major loan exhibitions of Victorian England in the generosity of private lenders and in the haphazard selection, eschewing thematic or chronological typology in favour of a more or less succulent artistic ragout.'[12] Given the diversity of the displays these are understandable responses, but to suggest that there were no systems of classification at work would be misleading. To the committee, curators, commentators and to a public that shared their mores and values, the exhibition formed a coherent programme of instructive lessons, spectacular demonstrations and self-congratulatory examples of cultural and commercial achievement. The degree to which the organisers identified themselves as authors of the programme was downplayed: 'the Curators and Committee have manifested an anxious desire to be impartial and have often sacrificed their personal feelings to a sense of what was due to the entertainment of the public and the success of the Exhibition'.[13] By suggesting that the only criteria for selection and display were based on a projection of what the public would enjoy, underlying ideological imperatives were obscured. In his analysis of nineteenth-century exhibitions the sociologist Tony Bennett argued that displays of industrial machinery and its products were deployed to communicate an ideology of progress to working people, in order to secure their complicit participation and obscure the extent to which their labour was exploited.[14]

Systems of classification manage and order what can be known. Philosophers and cultural theorists have emphasised the contingency of these systems, the ideologies embedded in them and their dependence on epistemological conditions.[15] The Picture Gallery, for instance,

[11] Kusamitsu, 'Great Exhibitions before 1851', p. 81.

[12] Giles Waterfield, *The People's Galleries* (New Haven, CT: Yale University Press, 2016), p. 59.

[13] 'Leeds Public Exhibition', *Leeds Mercury*, 13 July 1839, p. 5. The same source notes the principal curators of the exhibition: Fine Art—Mr J. Wilkinson and Mr T. Nunneley; Philosophical Apparatus—Mr Wm. West and Mr T.P. Teale; Models—Mr E. Turner and Mr P. Fairbairn; and Natural History—Dr P. Smith and Mr Calvert.

[14] Tony Bennett, 'The Exhibitionary Complex', in *The Nineteenth-Century Visual Culture Reader*, ed. by Vanessa R. Schwartz and Jeannene M. Przyblyski (London: Routledge, 2004), pp. 117–30 (p. 123).

[15] Michel Foucault, *The Order of Things: An Archaeology of the Human Sciences* (London: Routledge, 2001), p. xix.

aggregated what we would now understand to be discrete categories of objects:

> At the head of the room is a full-length portrait of his late Majesty William IV, in the robes of the Garter, by John Simpson; and at the opposite end is a full-length portrait of Queen Anne, by Sir Peter Lely... On tables down the centre of the gallery are many beautiful pieces of philosophical apparatus, such as telescopes, microscopes, astronomical instruments and air pumps... And in cases down the sides of the room, are collections of coins, medals, insects, manuscripts ... specimens of the stages of the woollen manufacture from the raw wool to the finished cloth.[16]

This spatial arrangement constructed a set of correspondences between the hanging of paintings on the walls, the display of philosophical apparatus on tables and more diverse selections of smaller objects and ephemera in cases. The fine arts were not distinguished from other kinds of commodities in the context. As the art historian Thomas Gretton has articulated, this conception of art could be quantified and put to work: it represented a set of skills, educational and moral principles and, above all, a means to increase the price of manufactured objects.[17]

In his encyclopaedic study of London's exhibitionary culture from the Elizabethan period to the middle of the nineteenth century, the literary historian Richard Altick suggested that the breadth of objects displayed in these polytechnic exhibitions was due in part to rapid socio-economic change: 'inventions, engineering models, scientific demonstrations were most in harmony with the temper of the moment. But the kinds of non mechanical manmade objects that had dominated the old-fashioned curio collection were far from losing their appeal.'[18] The art gallery, science museum and fairground were yet to emerge as specialised spaces, so it is important to further explore the ways in which these categories of objects were understood in their own time.

[16] 'Leeds Public Exhibition', *Leeds Mercury*, 13 July 1839, p. 5.

[17] Thomas Gretton, '"Art is cheaper and goes lower in France." The Language of the Parliamentary Select Committee Reports on the Arts and Principles of Design of 1835–1836', in *Art in Bourgeois Society, 1790–1850*, ed. by Andrew Hemingway and William Vaughan (Cambridge: Cambridge University Press, 1998), p. 88.

[18] Altick, *The Shows of London*, p. 390.

Paintings, works of art and curiosities

The collection of oil paintings and watercolours gathered together for display at the Leeds Public Exhibition was presented as a source of pride and a signifier of cultural achievement for the whole town.[19] Loans of paintings to exhibitions in Leeds were generally more prolific than comparable polytechnic displays, as a result of strong local private collections and the precedent set by the Northern Society for the Encouragement of the Fine Arts.[20] Although some works were loaned directly by the artists themselves, the majority were owned by the local aristocracy and the emerging middle-class collectors whose new wealth had been generated in Leeds.[21] The paintings were variously historical and contemporary, known and unknown, local and international. Although in some instances works of significantly contrasting cultural and economic value were hung alongside one another, there were a number of mechanisms employed to guide the visitor in their judgements (Figure 1.1).

The catalogue recorded that the majority of the paintings were displayed in same suite of three rooms on the first floor of the Music Hall that had been home to the exhibitions of the Northern Society in previous decades: the Ante-Room, the Picture Gallery and the Cabinet, with a small number displayed in the Orchestra. The Ante-Room contained the highest number of paintings for sale directly through the artist, alongside an almost equal number of specimens of natural history.[22] The Picture Gallery contained some of the most highly regarded works, including those attributed to Rubens, Rembrandt, Caravaggio, Correggio, van Dyck and Poussin.[23] The Cabinet had for the most part been reserved for the watercolours loaned by Francis Hawksworth Fawkes (1797–1871) of Farnley Hall.

The curators of all the disciplines represented in the exhibition were recognised for their contributions, but the curator of the paintings was singled out for particular praise:

> The hanging of the paintings, a matter of so much importance in order to their tasteful arrangement and their appropriate juxtaposition, has been most ably superintended by a talented artist,

19 West and Baines, *A Description*, p. 5.
20 Lomax, 'Buying Antiques in Early Victorian Leeds', p. 275.
21 Arscott, 'Without distinction of party', p. 139.
22 *Catalogue of the Exhibition*, pp. 1–6.
23 *Catalogue of the Exhibition*, pp. 6–17.

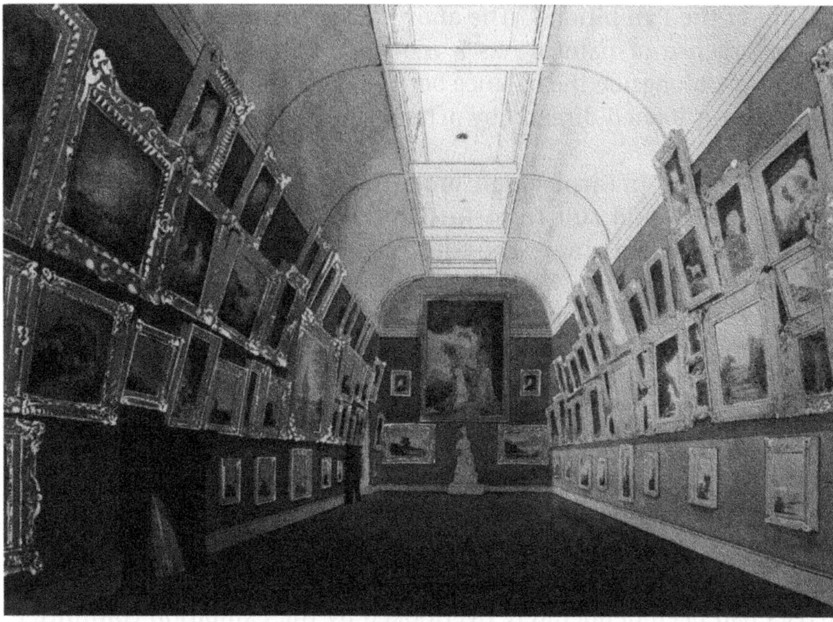

Figure 1.1 Joseph Rhodes, *Old Music Hall in Albion St Leeds (North Ante Room) in 1839*, 1839, watercolour, 45 × 35.5 cm. Leeds Museums and Galleries, UK/Bridgeman Images

> Mr. Bird, of London, to whom the lover of the fine arts will be under great obligations for much of their enjoyment in the picture gallery during the next three months.[24]

The language of this assessment points towards a perception that the selection and display of art required especially refined taste and specialised creativity, ideally cultivated in the metropolis. The status and prestige of the curator and lender were generally emphasised above the reputation of the artist. According to an anonymous correspondent to the *Art-Union* journal, the committee had not originally planned on the fine arts being a particularly prominent part of the exhibition. The correspondent wrote: 'About three weeks before the time appointed for opening, I made enquiry as to the number of pictures they had then promised. It appears they had no notion of paintings becoming a

[24] 'Leeds Public Exhibition', *Leeds Mercury*, 6 July 1839, p. 4.

feature of the Exhibition.'[25] The anonymous writer went on to recount the emergence of painting as a more substantial component of the exhibition, alongside the balance between pictures loaned by collectors and pictures loaned directly by artists:

> Mr. Fawkes, of Farnley Hall, broke the spell. He munificently offered all that would suit from Farnley Hall; consequently all Turner's inimitable drawings—the Fairfaxiana—portrait of Cromwell, and other interesting works were chosen. Then town and country seemed mad to send in their entire stock—to the dismay of the committee. They wisely chose a gentleman from London (an artist) to select, and reject; and, under the inspection of a committee, to hang, which I assure you, was no easy task. I have made the most careful investigation, and have to assure you that *only twenty-two* pictures, besides a very few portraits and smaller water-colour drawings, were *sent by artists* [original emphasis].[26]

The underlying purpose of the letter was to refute claims that living artists had been deliberately overlooked by the exhibition committee; the writer maintained that the committee had favoured the work of living artists above other loans.[27] The content of the correspondence also implied the existence of an undifferentiated abundance of works of art in local private collections alongside a generous and enthusiastic, but perhaps unsophisticated, approach to lending. That the exhibition committee chose an artist to curate this section was consistent with the prevailing belief that they were best placed to apply their training to the arrangement of works of art. Curators of the visual arts had yet to develop as specialised professionals, unlike their equivalents in the sciences who, as Waterfield recognised, operated as part of a longer curatorial tradition.[28]

Beyond the committee and their chosen curators, the descriptive guide foregrounded the donors to the exhibition, which provided an additional layer of public acknowledgement above the official catalogue and notices in the press. The most prestigious lenders were identified as Queen Victoria, who lent a portrait of herself by George Hayter (1792–1871); Francis Hawksworth Fawkes of Farnley Hall, who lent more than fifty pictures, including forty watercolours by J.M.W. Turner;

25 'Mechanics Exhibition', *Art-Union*, vol. 2, June 1840, p. 96.
26 'Mechanics Exhibition', p. 96.
27 'Mechanics Exhibition', p. 96.
28 Waterfield, *The People's Galleries*, pp. 17–18.

William Markham (1796–1852) of Becca Hall; George Lane-Fox (1793–1848) of Bramham Park; and Henry Lascelles of Harewood House (1767–1841), who also lent a portrait of himself by the celebrated painter and former president of the Royal Academy of Arts Sir Thomas Lawrence (1769–1830).[29] In his history of the emergence of the municipal art gallery in Britain, Waterfield argued that the monarchy's primary participation in, and promotion of, the visual arts was through its willingness to lend works of art to public exhibitions.[30] According to Waterfield this sense of shared cultural property—alongside the attendant obligation to make important works of art available for display—filtered down to the owners of private collections held in the nation's country houses. While the visual arts did not yet benefit from purpose-built public galleries as they did in parts of Europe, Britain's private collections of art were recognised for their international quality and significance.[31] Although lenders from the country gentry were recognised and celebrated, they were not the only demographic to donate objects to the exhibition, and it was not always the case that the aristocracy provided the fine art and the industrialists the machinery. In addition to his fifty pictures, Fawkes also lent a number of agricultural implements.[32] Industry and mechanical technologies were not confined to urban centres and neither were the gentry and the industrialist entirely separate—their commercial and cultural interests often intersected.[33]

The ways in which paintings were differentiated in this context were informed by utilitarian principles. Art objects were assessed according to their adherence to nature; a picture perceived as accurate became a document with the capacity to educate.[34] Both the subject and process of production could provide useful instruction.[35] The historian Amy Woodson-Boulton has argued that representational painting was interpreted as a means of accessing an imagined and transformative conception of nature and ideal beauty, which was thought to provide an experiential encounter with the potential to relieve the social and moral problems caused by living and working in industrialised urban centres.[36] Waterfield extended this idea to include the presentation of history,

29 West and Baines, *A Description*, p. 5.
30 Waterfield, *The People's Galleries*, p. 13.
31 Waterfield, *The People's Galleries*, p. 15.
32 'Leeds Public Exhibition', *Leeds Mercury*, 6 July 1839, p. 4.
33 Asa Briggs, *Victorian Cities* (London: Odhams, 1963), p. 151.
34 Arscott, 'Without distinction of party', pp. 149–50.
35 Altick, *The Shows of London*, p. 392.
36 Amy Woodson-Boulton, *Transformative Beauty: Art Museums in Industrial Britain* (Palo Alto, CA: Stanford University Press, 2012), pp. 2–8.

with audiences invited to treat paintings with historical subjects—whether narratives or portraits—as almost unmediated documentary evidence of the past.[37] The emphasis on fidelity to nature meant that even an artist as highly regarded as Turner was criticised according to these criteria. The reader of the descriptive guide was advised: 'The earlier works of Turner are superior to the later, the artist having unaccountably fallen into an unnatural and meretricious gaudiness of colouring... Turner's drawings are singularly true to nature in all its moods and appearances, and their combined brilliance and chasteness charm the eye.'[38]

These works by Turner shared the Cabinet with Fawkes's inherited collection of Fairfaxiana relating to the English Civil War, which included the swords of Sir Thomas Fairfax, Oliver Cromwell and General John Lambert in addition to items more akin to relics, such as 'a portion of his [Fairfax's] boot converted into a black-jack or drinking cup'.[39] The potentially divisive character of this collection was addressed in the descriptive guide, which balanced the commendable qualities of both the Royalists and the Parliamentarians: 'no person of right feeling can refuse pity to the fallen monarch, or deny to his partisans the merit of loyalty and self-devotion;—or to their more successful opponents the claim to military skill and courage, to sagacity, foresight, and resolution in their civil policy'.[40] These remarks reflected the sectarianism that pervaded the social, religious and political life of Leeds during this period, for which the collaborations generated by the exhibition were intended to provide a temporary truce if not complete reconciliation.[41]

The catalogue of the exhibition distinguished between painting and more generalised works of art. The categories of works of art and curiosities could intersect; the gender or ethnicity of the maker could shift the classification and interpretation of the object, and the colonial context cast objects as 'curious' by virtue of their perceived exoticism. Paintings by unnamed Chinese and Indian artists were displayed in the Picture Gallery—otherwise populated by works of British art—and discussed in terms that were at once admiring and patronising:

> No. 261, three miniatures executed at Canton by Lan-Qua, a Chinese artist, from well-known English paintings; No. 146, P.G.

[37] Waterfield, *The People's Galleries*, pp. 277, 280–81.
[38] West and Baines, *A Description*, p. 24.
[39] West and Baines, *A Description*, p. 31.
[40] West and Baines, *A Description*, pp. 30–31.
[41] Arscott, 'Without distinction of party', p. 138.

> [Picture Gallery] is still more curious, being a portrait of an English gentleman, by the same Chinese painter. The specimens seem to indicate a higher state of art than we are accustomed to suppose among the natives of that distant part of the globe. They are sent by Mr. Birtwhistle, surgeon, Cross Hills. In case, No. 263, P.G. are some paintings on talc, by a Hindoo [*sic*], contributed by the same gentleman; these, as faithful representations of the costumes and employments of our Indian fellow subjects, have a value which scarcely attaches to similar delineations by European draughtsmen.[42]

Similarly, the work of women—where it was identified as such—was characterised by its supposed femininity as opposed to its artistic or instructive merit. For example, 'the admirers of female ingenuity' were invited to 'compare the pursuits of modern ladies with those of a former period by an inspection of the Wax flowers in the Ante-Room'.[43] The 'miscellaneous curiosities' presented alongside the paintings and works of art were paradoxically more easily categorised; their sheer heterogeneity prevented them from entering other systems of knowledge. Numerous examples of this type of object could be cited from the catalogue. One of the most intriguing was item number 249 in the Picture Gallery: 'cinders from the combustion of 500,000 bank notes'.[44] Unfortunately the origin and purpose of this exhibit was not recorded.

Antiquities and natural history

In contrast to the paintings from private collections, antiquities and specimens of natural history could be borrowed from local collections with at least some form of established public access, the most prominent of which was the museum of the Leeds Philosophical and Literary Society at the nearby Philosophical Hall on Park Row.[45] John Calvert's museum on neighbouring Commercial Street was another important source. Calvert had amassed a collection of around 15,000 examples of 'biological, geological, and ethnological objects' over the course of thirty-five years. His museum opened in 1827 with a ticket price of one shilling, and he co-curated the specimens of natural history on display

[42] West and Baines, *A Description*, p. 25.
[43] West and Baines, *A Description*, p. 27.
[44] West and Baines, *A Description*, p. 27.
[45] West and Baines, *A Description*, p. 34.

at the 1839 exhibition alongside Dr P. Smith.[46] Just as works of art were dispersed across several rooms, objects categorised as natural history could be found in the Ante-Room, Picture Gallery and Saloon Gallery. The historian of science Samuel Alberti has shown that the subject and objects of natural science were increasingly popular during this period, which accounts for their prominence at the exhibition.[47] The Ante-Room contained mounted birds from Britain and more exotic species from Africa and the Americas.[48] In the Picture and Saloon galleries, the display was more varied and included insects, skulls and skeletons, fossils, shells and minerals.[49] The natural sciences were not only represented by static specimens; live birds were also presented to contrast with the objects of human culture on display: 'what a multitude of reflections suggested by the contrast between their present abode, amidst the tokens and results of the highest degree of civilization yet attained by man, and the solitude and wildness of their native forests!'[50]

The exhibition paired the natural sciences with antiquities. The rapid expansion of early Victorian infrastructure resulted in new discoveries in the local area. The construction of a new railway station inside the city walls of nearby York resulted in the excavation of archaeological material, which included 'Stucco composed of broken stones, bricks, and mortar, found in a Roman bath'.[51] Antiquities were categorised somewhat eccentrically under the following headings: Egyptian, Roman, Ancient Bricks, British and Carvings.[52] Ancient objects were not a major component of the 1839 exhibition; they would go on to have a greater role and presence at the Second Public Exhibition of 1843, which featured a designated 'Museum of Antiquities'.[53] This shortage was also expressed in the descriptive guide, which cited limited space and not the supply of objects as the determining factor.[54]

46 Edwin Kitson Clark, *The History of 100 Years of Life of the Leeds Philosophical and Literary Society* (Leeds: Jowett and Sowry, 1924), p. 38.

47 Samuel Alberti, 'Natural History and the Philosophical Societies of Late Victorian Yorkshire', *Archives of Natural History*, 30 (2003), 342–58 (p. 342).

48 *Catalogue of the Exhibition*, pp. 4–5.

49 *Catalogue of the Exhibition*, pp. 11–14.

50 West and Baines, *A Description*, p. 36.

51 West and Baines, *A Description*, p. 29.

52 West and Baines, *A Description*, p. 28.

53 Lomax, 'Buying Antiques in Early Victorian Leeds', p. 275.

54 West and Baines, *A Description*, p. 28.

Philosophical, chemical and electrical apparatus

The philosophical, chemical and electrical apparatus displayed at the exhibition would now be defined as more generalised scientific or technical instruments, although they represented distinct forms of knowledge and technologies during the nineteenth century. The boundary between the arts and the sciences was, however, more permeable in this context of public display. 'Philosophical Apparatus' included objects such as telescopes, orreries, surveying instruments and compasses.[55] The majority were to be found on the central tables in the Picture Gallery, with the powerful oxy-hydrogen microscope presented on its own in the Tuning Room and demonstrated at regular intervals.[56] Slide preparations were used with the microscope, showing 'insects, leaves, and other objects, among the most minute which the glass can reach, and also among the most wonderful and beautiful in their conformation, as well as living animalculæ in the most rapid motion, are shown prodigiously magnified'.[57] According to Altick, this technology 'contributed little to material progress but was well suited for show business'.[58] The display of 'Chemical Apparatus' was largely performative too in that it was the process that was considered more instructive than the object. These processes included distillation and the chemical analysis of atmospheric conditions.[59] The demonstration of a diving bell was also included under this category, during the course of which a rabbit, six mice and a lit lamp were submerged to illustrate the retention of air under water. Visitors were assured that 'the animals may be seen feeding, sporting, or at rest, unconscious of the strange element beneath which they are plunged'.[60]

The Saloon was the site of some of the most dramatic and spectacular aspects of the exhibition, and was said to resound 'with the noise of engines, machines, and scientific processes'.[61] Many of the experiments were selected for visual as well as scientific interest. Balloons were used to demonstrate the relative densities of air and gas: 'those exhibited are formed of animal membrane ... some are globular, others in the form of a fish. While new they ascend readily when filled with coal gas. One

[55] West and Baines, *A Description*, pp. 37–43.
[56] *Catalogue of the Exhibition*, pp. 14–17, 29.
[57] 'Leeds Public Exhibition', *Leeds Mercury*, 13 July 1839, p. 5.
[58] Altick, *The Shows of London*, p. 369.
[59] West and Baines, *A Description*, p. 44.
[60] West and Baines, *A Description*, p. 45.
[61] 'Leeds Public Exhibition', *Leeds Mercury*, 13 July 1839, p. 5.

may frequently be seen floating near the ceiling of the Saloon.'[62] The 'Electrical Apparatus' on display were similarly theatrical, promising 'Electric Shocks at intervals' and 'a Thunder House, for illustrating the effects of electricity on conductors to buildings, showing what would take place if struck by lightning when such a conductor was broken or damaged'.[63] How far these experiments communicated their scientific principles is subject to debate. Altick concluded: 'that the public was hungry for scientific knowledge proved irreconcilable with the public's stubborn insistence that, first of all, it be amused'.[64] Since the 1839 exhibition was the first of its kind in the town, it is likely that the committee based their selection and arrangement of scientific experiments on what had already proven popular with visitors to the National Gallery of Practical Science and the Polytechnic Institution in London.

Machinery, models, materials and manufacture

For those visitors whose labour was extracted through the factory system, the examples of industrial machinery and manufactures may have been the most familiar items on display. The descriptive guide offered the middle classes a framework for understanding these technologies: 'the rough and plain appearance inevitable with some of this description of specimens, is relieved, and their interest increased, by the taste with which they are disposed and exhibited'.[65] To further understanding of the machinery on display, it was often shown in full operation and the labour process was to some extent made visible, with the operative objectified as an archetypal example of their class. The descriptive guide highlighted the following example, in which the owners of the machine were identified but the operatives remained anonymous:

> The [card-making] Machine which in general first engages the attention of visiters [*sic*], is that for the manufacture of cards for wool or cotton... This machine belongs to Messrs. Curtis, Parr, and Walton of Sowerby Bridge; and in addition to lending it, they are

62 West and Baines, *A Description*, p. 45.

63 'Leeds Public Exhibition', *Leeds Mercury*, 13 July 1839, p. 5; West and Baines, *A Description*, p. 46.

64 Altick, *The Shows of London*, p. 363.

65 'Leeds Public Exhibition', *Leeds Mercury*, 13 July 1839, p. 5.

> at the whole expense of the workmen employed, and allow them to sell specimens of the work produced, the proceeds to be applied in aid of the funds of the Exhibition. None who can afford a few pence should leave the Saloon without a relic and memorial of labours evincing so much ingenuity.[66]

Similarly, the operative of the Jacquard loom—who was named as John Haywood in this instance—produced cloaks and waistcoats that could also be purchased by visitor as souvenirs.[67] In addition to examples of the mechanisation of textile production, a steel-cutting machine was demonstrated at regular intervals, with the dazzling sparks it generated described as 'splendid as well as curious'.[68] The terms used to describe the effects of this machine were very close to those used in relation to the scientific demonstrations; the process was presented as experimental, exciting and visually stimulating. The working machinery on display was positioned as both invention and entertainment. It sat between the celebration of technological progress represented by the scientific instruments on display and the staging of scheduled musical performances as collective experiences. This form allowed the mechanic and the operative to witness the (limited) celebration of their role, the consumer to better understand the processes of production, and the factory owner to present an image of public-spirited generosity and safe working conditions.

'For the benefit of the Mechanics' Institution'

The commentary surrounding the Leeds Public Exhibition tended to emphasise its educational benefits to the local population, but a primary motivating factor for staging the exhibition was to generate funds for the Leeds Mechanics' Institution. The annual report published by the Mechanics' Institution in 1833 revealed financial difficulties. In his contemporary account, the local minister and historian Edward Parsons (1797–1844) summarised the situation: 'the balance in the treasurer's hands, at the time when the committee commenced its operations, being less than any former period since the establishment of the Institution, it was thought essentially necessary to exercise that

66 West and Baines, *A Description*, p. 54.
67 West and Baines, *A Description*, p. 56.
68 West and Baines, *A Description*, p. 59.

economy in every department'.[69] The exhibition was intended to secure its financial position and extend its influence and activities. That the Institution subsisted in rented rooms was a source of embarrassment and did not compare well with other industrial centres with purpose-built Mechanics' Institutes. Without a permanent presence in the town, the Institution suffered a social invisibility that reflected badly on the committee, members and subscribers and, by extension, the wider population. Parsons' summary of the 1833 annual report articulated their aspirations:

> Although Leeds ... has not yet emulated the examples of some of the large and wealthy towns in the manufacturing districts, it is by no means inferior to any of them in scientific taste and acquirement, and we are certain that the munificence of its inhabitants will soon place its Mechanics' Institution, both in external appearance and internal convenience, on a par with any other edifice reared for a similar purpose in the kingdom.[70]

The potent combination of flattery, reprimand and civic rivalry inflected the coverage of the exhibition too. In a direct address to 'the Tradesmen and Mechanics of Leeds' through the *Leeds Mercury*, Edward Baines Junior urged the manufacturing and operative classes to continue their support of the Mechanics' Institution after the closure of the exhibition and referred to the former as 'the seed shop' and the latter as 'the garden'. His appeal was not directed towards individual improvement so much as collective social progress: 'if in any town a Mechanics' Institution ought to be largely supported, it is in Leeds; because the prosperity, it may almost be said the existence, of our great population is dependent on their proficiency'.[71] The perceived connection between proficiency and prosperity had been embedded in the aims of the Institution from its inception, when it was confidently declared: 'the scientific instruction which will give a more thorough knowledge of their arts, will greatly tend to improve the skill and practice of those

69 Edward Parsons, *The Civil, Ecclesiastical, Literary, Commercial, and Miscellaneous History of Leeds, Bradford, Wakefield, Dewsbury, Otley, and the district within ten miles of Leeds* (Leeds: F. Hobson, 1834), p. 104.

70 Parsons, *History of Leeds*, p. 105.

71 Edward Baines Junior, 'To the Tradesmen and Mechanics of Leeds, on the Public Exhibition and the Mechanics' Institution', *Leeds Mercury*, 28 September 1839, p. 5.

classes of men who are essentially conducive to the prosperity of this large manufacturing town'.[72]

In financial terms the Leeds Public Exhibition did benefit the Mechanics' Institution. The diarist John Mayhall recorded the profit as approximately £1,780 after expenses.[73] In a letter to the *Art-Union* journal in May 1840, the anonymous writer alleged that the displays of fine art and other static objects were not the primary source of the profits, but rather: 'It was the steel-cutting, the carding-machine, and a thousand others; the *glee singing*, the band of music; but more especially the *evening fair*, that brought the money [original emphasis].'[74] The italics implied that it was the support of the working classes that really generated the profit, although it is not possible to evidence this with any precision. Whatever the proportion of income brought by different parts of the exhibition, the resulting profit was sufficient for the Institution to purchase its own building, although it was not enough to fund the construction of new premises as had been hoped.[75] A site was chosen at 12 South Parade, close to prestigious Park Row and the Philosophical Hall, which was purchased by the Institution in February 1840 and officially opened in November 1841.[76] Built as Walton's Music Saloon in 1837, 12 South Parade had been quickly appropriated by Owenite socialists who established a 'Socialist Hall' there for lectures and discussions.[77] The eviction of this group through the sale of the

72 *The Leeds Institute of Science, Art and Literature: An Historical Sketch of One Hundred Years' Work, 1824–1923* (Leeds: Jowett and Sowry, 1923), p. 6.

73 John Mayhall, *The Annals and History of Leeds, and other places in the County of York, from the earliest period to the present time* (Leeds: Joseph Johnson, 1860), pp. 461–62.

74 'Mechanics Exhibition. To the Editor of "The Art-Union"', *Art-Union*, vol. 2, June 1840, p. 96.

75 The Leeds Mechanics' Institution succeeded in building its own premises after merging with the Leeds Literary Institution in 1842. This amalgamation was named the Leeds Institute of Science, Art and Literature. The new Institute was designed by the architect Cuthbert Brodrick and built between 1865 and 1868.

76 *The Leeds Institute of Science, Art and Literature: An Historical Sketch of One Hundred Years' Work*, p. 7.

77 In 1838 the 'Leeds District Board of the Association of All Classes of All Nations', which listed its office at Walton's Music Saloon, published a pamphlet that refuted claims that Robert Owen's New Lanark had been a failure, by reproducing a report written by Edward Baines, Robert Oastler and John Cawood in 1819. See also J.F.C. Harrison, *Learning and Living 1790–1960: A Study in the History of the English Adult Education Movement* (London: Routledge, 1961), p. 110; R.J. Morris, *Class, Sect and Party: The*

building marked its transfer from a space associated with a radical vision of working-class emancipation through Chartist activism to a much 'safer' form of working-class improvement essentially prescribed by a middle-class committee.

A self-interested philanthropy

The Leeds Public Exhibition offered a very prominent platform for the projection of intellectual refinement, cultivated taste and public spirit. The primary vehicle for this publicity was the catalogue, which listed objects by subject, artist—where known and appropriate—and contributor. The publication of the names of contributors ensured that both individuals and the collective achievement of the town were recorded and communicated.[78] In addition to the prestige of being recognised as a lender, there were financial gains to be made too. At its most transparent, items marked in the catalogue with an asterisk were for sale. These objects were generally oil paintings and watercolours submitted and sold by the artist as a continuation of the practice established by the annual exhibitions of the Northern Society for the Encouragement of the Fine Arts. There were also more esoteric objects for sale, such as item number 295a: 'Stromeyer's Splint for deformity of the foot, on a new and much improved principle', and item 316: 'Marsh's Apparatus for Detecting Arsenic', both of which were on display in the Picture Gallery.[79] The exhibition was used to advertise industrial machinery for sale too, where sellers benefited from being able to demonstrate their wares in full operation. One example appeared in the 'Sales by Private Contract' section of the *Leeds Mercury*, which advertised a four horsepower high-pressure engine, 'at present Working at the Leeds Public Exhibition', available for sale via James N. Greenwood, a whitesmith of Wharf Street.[80] Although it is not possible to determine whether or not the advertisement generated a sale, that the exhibition was used as a shop window by local artists and manufacturers was

Making of the British Middle Class, Leeds 1820–1850 (Manchester: Manchester University Press, 1990), pp. 310–11; and *The Leeds Institute of Science, Art, and Literature: Historical Sketch, 1824–1900* (Leeds: Goodall and Suddick, 1901), p. 5. After the Leeds Mechanics' Institution and Literary Society moved into purpose-built premises in 1862, 12 South Parade housed the Young Men's Christian Association and was demolished in 1904.

78 Arscott, 'Without distinction of party', p. 139.

79 *Catalogue of the Exhibition*, p. 15.

80 'Sales by Private Contract', *Leeds Mercury*, 26 October 1839, p. 7.

significant because it shows how embedded commerce was to this form of polytechnic exhibition. Their economic, social and cultural aims were seen to reinforce one another, rather than remaining mutually exclusive. Unlike the discourse surrounding the galleries and museums that would emerge in the second half of the century, there was no need to even claim that the market was not in operation in this sphere.

Even if an item were not for sale, there remained an economic logic for making an object or collection visible to a wider audience. The language of the newspaper coverage emphasised the egalitarian, altruistic and munificent qualities of the donors: 'the walls of splendid mansions have been stripped of the choicest productions of genius, and cabinets have been unlocked, and the costly contents emptied out, in order that a rich intellectual feast may be spread before our vast population'.[81] Aside from the prestige of cultural philanthropy, the public display of an object bolstered value through the construction of a context that signified its authenticity and cultural importance.

In addition to the self-interested philanthropy displayed by lenders, it was also in the interests of the owners of the means of production to support the exhibition. This support came in two forms: from the loan of industrial machinery and through the attendance of their workforce. Edward Baines Junior saw the Mechanics' Institution in general and the exhibition in particular as a means of maintaining and improving the economic position of the region when he issued these words of caution: 'the manufacturers of Leeds are embarked in an arduous competition. The markets of the whole world consume our products; and the manufacturers of many countries of Europe and America are incessantly striving to gain the superiority over us. Defeat, then, would be ruin.'[82] The exhibition was part of the apparatus that worked against the alienation of the wage labourer by offering a celebratory and even heroic narrative of their place in the world and entertaining distractions from their working life.[83] The Leeds Public Exhibition at least partially functioned as a mechanism to maintain social, political and economic stability.[84]

As the architectural vehicle for the exhibition, the Albion Street Music Hall was not a neutral space. The Music Hall was built between 1792 and 1794 and closed in 1870. A description published five years before the 1839 exhibition detailed its various uses:

81 'Leeds Public Exhibition', *Leeds Mercury*, 6 July 1839, p. 4.

82 Baines, 'To the Tradesmen and Mechanics of Leeds', p. 5.

83 Altick, *The Shows of London*, p. 368.

84 Morris, *Class, Sect and Party*, p. 319.

> The ground floor was for some years occupied as a hall for woollen manufacturers, especially for blankets, and afforded accommodation to those clothiers who were excluded from the Cloth Halls. It received, and for some time retained, the ignominious appellation of Tom Paine's Hall. It is now appropriated to other purposes. The Leeds Concerts have long been conducted with great spirit and considerable success; the hall however has frequently witnessed exhibitions of a far more impressive character than its musical assemblies; it has often formed the scene in which the claims of the noblest institutions of British Christianity and benevolence have been presented to the consideration and the ever ready liberality of the inhabitants of the town.[85]

The Music Hall was a site of overlapping architectural, commercial, cultural and religious meaning. The exhibitions referred to above were staged by the Northern Society for the Encouragement of the Fine Arts, established in 1808 with the aim of promoting the 'British School' of artists.[86] The Northern Society held a number of exhibitions at the Music Hall from 1809, in a suite of three rooms on the first floor.[87] They were given permission to adapt the interior by adding a gallery in 1823 and lighting the saloon from the roof by 1825.[88] The historian R.J. Morris argued that these exhibitions provided a model that was specific to Leeds; by the 1830s the town had an established culture that did not rely on the precedents set by other towns and cities. They primed the local middle classes to value the art objects presented by the polytechnic exhibitions of the late 1830s and early 1840s.[89] The art historian and curator Giles Waterfield identified the importance of the 'old families' of Leeds in the development of this culture of collecting and display.[90] The local gentry—in collaboration with those whose wealth had been made more recently in manufacturing, trade and finance—formed an 'unexpectedly sophisticated' social, economic and political elite who dominated exhibition making in Leeds for the rest of the century.[91]

85 Parsons, *History of Leeds*, p. 136.
86 Trevor Fawcett, *The Rise of English Provincial Art: Artists, Patrons, and Institutions outside London, 1800–1830* (Oxford: Clarendon Press, 1974), p. 168.
87 The Picture Gallery was the largest of the three rooms at 60 feet in length, with the Cabinet at one end and the Ante-Room at the other.
88 Fawcett, *The Rise of English Provincial Art*, p. 114.
89 Morris, 'Leeds and the Crystal Palace', p. 285.
90 Waterfield, *The People's Galleries*, p. 56.
91 Waterfield, *The People's Galleries*, p. 56.

Although the objects and their lenders projected a prestigious picture, the Music Hall retained traces of more complicated signifiers. Also known as the 'Irregulars' Cloth Hall', from 1793 the lower floor had operated as a space to trade for those who had not completed their seven-year apprenticeship.[92] Modest in size and 'unpretentious in the way of architecture', the Music Hall was decorated simply and could accommodate audiences of up to around 1,000 people to see musicians such as the virtuoso Italian violinist Niccolò Paganini (1782–1840) perform, as he did in 1832.[93]

The route the visitor took through the Music Hall at the exhibition of 1839 was tightly prescribed and did not permit visitors to wander freely through the rooms. The *Leeds Mercury* noted that 'the arrangements for preserving order are good, and have proved very effectual during the fair days. One of the regulations, which is needful to prevent confusion, is that the visitors shall move onwards through the suite of rooms, and not turn back from one room to another.'[94] The one-way circulation was reinforced through the official catalogue, which gave the order of the rooms and also, in some cases, the order in which the walls, display cases and individual objects should be viewed. The route took the visitor directly to the first floor, through the interconnecting Ante-Room, Picture Gallery and Cabinet, to the largest space containing the Saloon and Orchestra, to the Tuning Room and Saloon Gallery, and finally downstairs to the ground-floor Lobby.[95] It was the body of the working-class visitor that was most explicitly subject to order. The *Leeds Mercury* published a notice from the exhibition committee explicitly addressed to the working classes: 'Operatives are requested not to come in their working dresses.' The notice continued:

> The reasonableness of this request will be acknowledged by all. About ten persons arrived one day this week from a considerable distance to see the Exhibition, but their dresses were so likely to soil whatever they came in contact with, that the committee were reluctantly obliged to ask them to postpone their pleasure till another day.[96]

92 Kevin Grady, 'Commercial, Marketing and Retail Amenities, 1700–1914', in *A History of Modern Leeds*, ed. by Derek Fraser (Manchester: Manchester University Press, 1980), pp. 177–99 (p. 179).

93 George Haddock, *Some Early Musical Recollections of G. Haddock* (London: Schott, 1906), pp. 37–38.

94 'Leeds Public Exhibition', *Leeds Mercury*, 13 July 1839, p. 5.

95 *Catalogue of the Exhibition.*

96 'Leeds Public Exhibition', *Leeds Mercury*, 20 July 1839, p. 5.

The rhetoric of the purpose of the exhibition, its relatively accessible ticket price and long opening hours appeared to actively include working-class visitors. It is clear from this example, however, that if visitors were not able to present themselves in a way that was acceptable to the organisers, through the standard of their clothing, cleanliness and behaviour, they could be denied admittance to the exhibition. That the visitors referred to above had travelled 'from a considerable distance', presumably directly from their place of work, suggests a commendable level of commitment, which was not matched by those who simply viewed them as potential pollutants.

Social class was also a factor in the accessibility of the exhibition to women. The Albion Street Music Hall was located in the ward of Mill Hill, which had a high population of women as a result of the affluence of the area. Wealthy women tended to live longer and their families employed more women as servants.[97] The domestic sphere was positioned as inherently feminine and the public sphere, inherently masculine. Historians such as Simon Gunn have pointed towards the gendered space of the urban sphere and the emergence of the figure of the 'public man' during this period, who came to dominate political discourse.[98] As the art historian Dana Arnold has noted, a 'public woman' had been synonymous with prostitution until the early nineteenth century, when it became more acceptable for women of a higher social class to participate in social and cultural activities outside the home.[99]

The local press considered the presence of women at the Leeds Public Exhibition noteworthy. The *Leeds Mercury* reported: 'a large party of young women who were employed by the Mayor of Leeds, and whose appearance was very respectable, visited the Exhibition, a holiday and tickets having been given them by the gentleman'.[100] It was only the appearance of women and working-class visitors that was subject to comment in this way. In contrast to the group that had been prohibited from the exhibition as a result of their clothing, the 'respectability' of these women was conferred by their employer. The mayor, James Holdforth, was the owner of silk mills in Leeds and the chairman of the committee that organised the exhibition. Alongside other local employers, Holdforth provided his 'principal men and mechanics'

97 Morris, *Class, Sect and Party*, p. 39.

98 Simon Gunn, *The Public Culture of the Victorian Middle Class: Ritual and Authority and the English Industrial City 1840–1914* (Manchester: Manchester University Press, 2000), p. 15.

99 Dana Arnold, *Re-presenting the Metropolis: Architecture, Urban Experience and Social Life in London 1800–1840* (Aldershot: Ashgate, 2000), pp. 26–27.

100 'Leeds Public Exhibition', *Leeds Mercury*, 10 August 1839, p. 5.

with season tickets.[101] During the last month of the exhibition, free or discounted admission was granted to other groups too, including soldiers stationed in barracks and children who attended Lancastrian and National schools.[102] The distinction between those provided with season tickets to visit the exhibition at their leisure and those issued with a specific time to attend indicated a variation in the levels of access granted to a stratified working-class public. The time of day that the visitor was able to attend determined their proximity to the objects on display and the experience of moving through the space. Between 7 p.m. and 10 p.m. the rooms were described as almost unbearably crowded.[103] For those restricted to visiting outside normal working hours, the exhibition did not necessarily function as a space of respite from the conditions of living and working in an industrial town.

Conclusion

The Leeds Public Exhibition of 1839 was a product of both general and particular conditions. Comparisons can be drawn with national institutions and exhibitions while recognising what was specific and distinct about its context and execution. As the historian Asa Briggs has shown, civic pride and local identity could be stimulated through the rhetoric of regional rivalry, coupled with a more cooperative position towards attempting to undermine London's cultural and political hegemony. Where Briggs and others have concentrated their analysis on architecture, infrastructure and civic rituals as the means by which urban centres attempted to assert their superiority, throughout this book we will witness the ways in which temporary public exhibitions—and later permanent civic collections, art galleries and museums—operated as an index of Leeds's sense of its place in the world.[104] The 1839 exhibition brought together objects that were not readily accessible to the public and enabled an element of class tourism that operated in two directions: the working classes were able to access—albeit in a restricted way—objects of high culture, and the middle classes could experience a sanitised representation of the factory floor with its machines and operatives, noise and motion. The objects of art, the sciences and industry were not placed in opposition:

[101] 'Leeds Public Exhibition', p. 5.
[102] 'Leeds Public Exhibition', *Leeds Mercury*, 14 September 1839, p. 5.
[103] 'Leeds Public Exhibition', *Leeds Mercury*, 20 July 1839, p. 5.
[104] Briggs, *Victorian Cities*, p. 85.

they were presented with equal instructive weight and unified through the discourses of taste and progress.

The development of Leeds as a significant centre of industrialised production had an impact on the objects selected for display and demonstration, the ways in which they were interpreted and the people who could access them. In some cases the employers of wage labourers took part in a form of social engineering to control the leisure activities of their workforce, a practice that connected exposure to culture with temperance and self-regulation. The culture that the working classes were given access to was a construction of the Leeds Philosophical and Literary Society and the Leeds Mechanics' Institution. These voluntary societies provided a fragile and temporary unity to an otherwise diverse middle-class experience; they were essential to the development of a coherent identity during a period of religious and political sectarianism. The Leeds Public Exhibition allowed the local aristocracy to maintain their position as custodians of heritage and continuity, while the middle classes laid their claim to progress and public culture.

2

The Second Leeds Public Exhibition, 1843

One of the first indications that a second exhibition might be held in Leeds on the model of the 1839 exhibition came in January 1843, when the *Leeds Mercury* reported that at the annual meeting of the Leeds Mechanics' Institution and Literary Society, it was suggested that 'a second exhibition in this town might be tried with a reasonable prospect of success'.[1] The suggestion was consolidated in public the following month, when Edward Baines Junior, in his capacity as vice-president of the Mechanics' Institution, announced the probability of another exhibition being held in the summer of that year.[2] By the end of February a committee had been formed to organise the exhibition, which was again intended to financially benefit the Mechanics' Institution. The committee was to include men from both within and beyond the Institution, alongside a number of the same people connected with the 1839 exhibition. The committee confidently declared that the new iteration would be 'even more splendid and interesting than that held four years since'.[3]

While the *Leeds Mercury* reported on these early developments with a sense of confident expectation, the *Leeds Times* struck a note of caution alongside its similarly effusive description of the projected benefits of the project:

> We trust that the same active spirit and benevolent zeal, which on a former occasion produced such great and beneficial results, will again be manifested on this occasion; and though, from the

[1] 'Annual Meeting of the Leeds Mechanics' Institution and Literary Society', *Leeds Mercury*, 21 January 1843, p. 5.

[2] 'Leeds Mechanics' Institution and Literary Society', *Leeds Mercury*, 18 February 1843, p. 5.

[3] 'Intended Exhibition in Leeds', *Leeds Mercury*, 25 February 1843, p. 5.

> altered circumstances of the times, the same large increase of success, in a pecuniary point of view, may not crown these efforts, the example of other towns, where similar Exhibitions have very recently taken place, or are even now in progress, affords sufficient encouragement, that with due zeal and energy on the part of the managers, and a liberal spirit on the part of contributors, very beneficial results may be produced.[4]

The *Leeds Times* recognised that the economic context had already changed since 1839 and that even the greatest efforts might not result in the same returns. Its faith in the endeavour was, however, supported by the success of the 1839 exhibition and comparable exhibitions held since in other towns and cities. The question of civic rivalry and competition quickly emerged again, with Hull, Liverpool and Manchester noted as having recently enjoyed successful exhibitions.[5] It was no longer enough for Leeds to have presented a single successful exhibition four years ago. In order to claim its position—real or imagined—as a leading town in the production of both capital and culture, it had to demonstrate its capacity not only to match what was achieved in 1839, but to substantially improve on it, in the spirit of progress and innovation. The committee reaffirmed their belief in the social function of these exhibitions through an advertisement in the *Leeds Mercury* on 1 April 1843:

> The Committee hold it to be a component and consistent part of the various endeavours to benefit our age, to give to the People, in Exhibitions like the one contemplated, the opportunity of inspecting and profiting by Works of Nature and of Art, and the Productions of Science, which otherwise they would never have beheld, and which could in no other way be seen in one place even by the most wealthy. Such Exhibitions are universally admitted to improve the taste and inform the mind.[6]

The committee positioned their activities as part of a broader social project, characterised by benevolent cultural and educational activities directed towards 'the People'—a group who were not consistently defined. Through this statement, it appears that the committee thought

4 'Second Public Exhibition of the Mechanics' Institute', *Leeds Times*, 25 February 1843, p. 5.

5 'Leeds Second Public Exhibition', *Leeds Mercury*, 1 April 1843, p. 1.

6 'Leeds Second Public Exhibition', p. 1.

the primary benefit to the visitor would be an improved sense of judgement, which we might reasonably extrapolate to mean a set of behaviours and values more recognisable to the middle classes responsible for organising these events. In the minds of the committee, the ideal working-class visitor would emerge from their encounter informed and ready to produce, consume and act according to a set of principles that would otherwise not be available to them through education and travel.

Soliciting donations

The committee took the opportunity on 1 April to begin the process of soliciting donations of objects to be displayed for a period of three months. They asked for the loan of 'valuable Specimens of Arts, Models of Machinery, &c.', and made it clear that lenders would be spared the expense of transporting their donations and indemnified against damage and 'all articles made of fragile materials, or of delicate texture, will be exhibited under glass cases or shades; and experienced and careful Workmen will be employed in hanging and taking down the Paintings'. Slightly less reassuring, however, was the admission that, 'in the former Exhibition very few articles were injured, and those only to a very slight extent'.[7] Artists and manufacturers were also invited to send their own productions and to list them for sale, making it a commercial opportunity not just for the Mechanics' Institution but for individuals and businesses.[8] The advertisement went into a little more detail about how exactly the expected proceeds would benefit the Mechanics' Institution: 'Any surplus which may arise from the proceeds of the Exhibition will be devoted towards liquidating the debt still resting on the Hall of the Leeds Mechanics' Institution and Literary Society, which building was mainly purchased with the proceeds of the former Exhibition.'[9] Not only did the profits from the 1839 exhibition fall short of enabling the Mechanics' Institution to realise a purpose-built headquarters, they had also been insufficient to purchase the building at 12 South Parade outright.

An article in the same issue of the *Leeds Mercury* called attention to the advertisement on its front page and directed its appeal for loans to 'the nobility and gentry generally, and particularly those connected

7 'Leeds Second Public Exhibition', p. 1.
8 'Leeds Second Public Exhibition', p. 1.
9 'Leeds Second Public Exhibition', p. 1.

with Yorkshire', and went on to reassure these potential donors that 'the Committee comprises the names of many of the principal inhabitants of the town, without distinction of party,—men of all parties feeling anxious to promote an object which cannot fail to be attended by great public advantage'.[10] While the exhibition itself was notionally directed towards all social classes, it is clear that at this stage, only those from the middle and upper classes were being actively invited to contribute objects. As such, those with wealth, influence and collections of valuable objects (in both cultural and economic terms) could shape the narrative of the exhibition. This situation was confirmed when it was reported that Francis Hawksworth Fawkes of Farnley Hall, 'the largest contributor to the former Exhibition in this town, has most liberally and freely declared his readiness to contribute from his valuable gallery of paintings and his other curiosities to the Exhibition'.[11]

By the beginning of June the committee were beginning to consolidate some of the arrangements for the forthcoming exhibition, which would again take place at the Albion Street Music Hall. It was proposed that the exhibition open to the public in mid-July, and with six weeks to go, the committee celebrated the 'offers of loans of specimens in works of art, nature, history, machinery and models, antiquities, materials and products of manufacture, and philosophical apparatus ... of the most interesting character', while continuing to request donations. The ambition of the exhibition seemed to be somewhat tempered by the statement that the committee were 'endeavouring to make the forthcoming one at least as attractive as the former one'—an exercise in managing expectations quite unlike earlier hyperbolic pronouncements.[12]

Two weeks later the *Leeds Times* followed with a much more optimistic assessment of the preparations, which it described as 'proceeding most favourably' as a result of the 'daily receipt of offers of first-class works of art, not only from the inhabitants of the town, but also from the nobility and gentry of the neighbourhood'.[13] It was reported that a number of valuable paintings had been received and, further, that new lenders had their concerns assuaged about lending works of fine art because the steam engines and working models would be displayed separately in an adjoining building, unlike their proximity at the previous exhibition.[14]

10 'Leeds Second Public Exhibition', *Leeds Mercury*, 1 April 1843, p. 5.
11 'The Intended Public Exhibition', *Leeds Mercury*, 22 April 1843, p. 4.
12 'Intended Public Exhibition at Leeds', *Leeds Mercury*, 3 June 1843, p. 5.
13 'Leeds Public Exhibition', *Leeds Times*, 17 June 1843, p. 5.
14 'Leeds Public Exhibition', p. 5.

Would this separation be analogous to the anxieties surrounding the mixing of social classes? It is difficult not to draw a parallel between the dirt and noise caused by the working machinery and the perceived dirt and noise brought by working people. The following week, the *Leeds Times* renewed calls for donations, concluding with their belief that 'the importance of these Exhibitions in creating and fostering a taste for rational amusements, is too obvious, to need our insisting upon it'.[15] This statement, so close to using an argument of 'common sense' to avoid analysis or explanation, lays bare the operation of ideology through these exhibitions. The term 'rational'—most often used as part of the phrase 'rational recreation' during this period—stands in for the idea that if only working men could avoid the irrational temptations of alcohol, gambling and other activities offensive to the Victorian drive towards public order, they could operate as stable, reliable and productive units with no need to look to the state for aid and no desire for revolution.

On 1 July, the month during which the exhibition was due to open on a yet unspecified date, both the *Leeds Mercury* and the *Leeds Times* ran the same short article calling once again for donations. It is worth reproducing in full for the strategies it employed and the shift in language from earlier appeals:

> We again beg to draw the attention of our readers to the advertisement which appears in our columns, requesting objects suitable for the purpose of the Exhibition. Many will remember the pleasure the attractions of the one held in the town four years ago. At that time, persons of all ranks crowded to the rich collection which was open to them; and we may add, that all were astonished that so large a selection of interesting objects could be got together, from the town and neighbourhood. Indeed Leeds rivalled for a few months, those cities of the continent, on whose public galleries travellers are never tired of expatiating. We earnestly solicit that those who are in possession of articles of virtue, or of antiquarian interest, or of Natural History, or others which are valuable for their rarity, will send in their names to the secretaries. This is the only way in which the Exhibition can be made worthy of the reputation of the town, or really conducive to the object for which it is intended.[16]

[15] 'Public Exhibition', *Leeds Times*, 24 June 1843, p. 8.

[16] 'Leeds Public Exhibition', *Leeds Times*, 1 July 1843, p. 5; 'Leeds Public Exhibition', *Leeds Mercury*, 1 July 1843, p. 5.

First the article requested support on the basis of the success of the 1839 exhibition and emphasised that it was open to 'all ranks' of people. It celebrated the collections that were brought together for the occasion and argued that, as a combined effort, it allowed the town to compete with the cultural clout of Europe, if only for a short time. As the committee specifically requested objects of *virtù*, natural history and other rare objects, it suggests that by this point they had sufficient numbers of works of fine art and manufacturing of sufficient quality to meet or exceed the expectations set by the first exhibition. The final two points of the article seem precision-engineered to generate both guilt and a sense of civic responsibility. It communicated the message that if they did not contribute, other people, other towns, would think Leeds not worthy of its reputation; the exhibition would fail to raise a profit for the Mechanics' Institution and 'the people' would not benefit from the improvement of their taste, morals, knowledge and behaviour.

Installation and promotion

The committee began to install the exhibits during the last week of June 1843, with an organ and a tank to accommodate a diving bell—the latter having also been a popular feature of the 1839 exhibition on a smaller scale. Porters were poised to collect the objects promised to the exhibition in the second week of July.[17] The first advertisement was published on 15 July, which publicised the opening date for the first time: Wednesday 26 July 1843.[18] The *Leeds Mercury* reported that the attractions were 'as various as they are interesting, and many of them entirely different in character from those exhibited in this town four years ago... In the department of fine arts, especially, we are assured, a very rich treat will be afforded.'[19] The promotion had to strike a delicate balance between reassuring potential visitors that the experience they enjoyed in 1839 would be repeated, but not so closely that there would be nothing new to see or experience. What seems extraordinary now is that just eleven days before the public opening, the committee continued to request donations for display. The timing of the flow of objects before—and even during—the period of public display will be a point to which we will return, as it demonstrates that what constituted

[17] 'Leeds Public Exhibition', *Leeds Times*, 1 July 1843, p. 4.

[18] 'Leeds Second Public Exhibition', *Leeds Mercury*, 15 July 1843, p. 1.

[19] 'Leeds Second Public Exhibition', p. 5.

'The Second Public Exhibition' was never quite as static as the published catalogue with its numbered list of objects might suggest.

The context of competition had also changed since 1839, with an increase in touring attractions that used some of the same spectacles to draw in the crowds. For instance, alongside an advertisement placed four days before the opening, another advertisement informed readers of an extended run of a display to rival the exhibition, only moments away on the next street. It boasted:

> Immense success of the Grand Pictorial Exhibition, in the large room of the Commercial Buildings, Mr. Stephens, from the Royal Polytechnic Institution, London, will exhibit, for one week longer, his gigantic and magnificent dissolving diorama, with an entire change of scenery, consisting of upwards of 150 views of Great Britain, India, China, Egypt, Syria, and the Holy Land, each view covering an area of 500 square feet, and exhibited by the aid of the Oxy-Hydrogen Lime Light. After which will be disclosed the awful secrets of the Eleusinian Mysteries; to conclude with the Ombres Fantastiques, being a Series of the Most Humorous Designs ever produced. First exhibition to commence at Six o'clock; Second ditto at Eight. Admission 6d.,—reserved seats 1s. Children and Schools half-price. N.B. Mr. S. begs to inform the Gentry and Inhabitants of Leeds that at the particular request of several influential individuals, there will be a select and fashionable exhibition, at 3, p.m., commencing on Friday, the 21st July.[20]

It is possible that the duration of this exhibition was deliberately extended to coincide with the opening of the Second Public Exhibition in order to capitalise on an anticipated increase in potential visitors to both displays. The event emphasised its origins in a London institution and, more specifically, an institution that had provided one of the principal models for the Leeds exhibitions of 1839 and 1843. It too combined the scientific with the spectacular and encouraged repeat visits through revolving displays. At 6d. for admission, it matched the cost of a single ticket to its rival across the street. It is important to note that the ticket price had not been inflated since the 1839 admission and remained the same for season tickets too, at 2s. 6d.[21]

[20] Advertisement and Notices, *Leeds Times*, 22 July 1843, p. 1.

[21] 'Leeds Second Public Exhibition', *Leeds Mercury*, 22 July 1843, p. 1.

Previews and reviews

The *Leeds Mercury* continued its strong support of the exhibition and whetted the appetites of its readers with a short preview on 22 July 1843, which promised that 'a high treat is in store... In the department of the fine arts it will be, beyond all doubt, the finest collection ever exhibited in this town.'[22] It provided a full assessment on 29 July 1843, on the day that the exhibition opened (three days later than first advertised) to the public. The unnamed reporter was typically effusive about the presentation: 'Having still a most lively recollection of the beauty and excellence of the Exhibition held in Leeds four years ago, we hesitate not in saying that this present one will carry away the palm from its competitor.'[23] Not only was Leeds in competition with its regional and national rivals, it was in competition with its own past, and if this account is to be believed, it was winning. The description of the contributors was marginally more inclusive than the 'nobility and gentry' targeted for their treasures, and the language was closely aligned to the model of the voluntary society in its ambition to bring together different social classes and political complexions:

> Men of all ranks and parties have entered cordially into the undertaking. The wealthy have sent the treasures of their galleries and cabinets,—the antiquarian has parted for a time with that he holds most dear,—men of science have given up the precious contents of their museums and their still more precious time,—the humble mechanic has laboured unweariedly to present a model or specimen in some branch of the useful arts. Noblemen and gentlemen far and near have most generously responded to the appeal made to them.[24]

There is a distinction here between the useful and beautiful, and an indication that the labour of the 'higher' ranks has been largely emotional: the painful, though temporary, parting with their precious collections, rather than the manual labour of the 'humble mechanic' who fashioned their own contribution. The paragraph quoted above is repeated in different ways across the reporting of the exhibition, in the way that it established both unity and difference between the social classes and their economic power. In addition to the lenders,

22 'Leeds Second Public Exhibition', p. 5.

23 'Leeds Second Public Exhibition', *Leeds Mercury*, 29 July 1843, p. 5.

24 'Leeds Second Public Exhibition', p. 5.

the article went on to describe the range of potential audiences in the following terms:

> And there has a collection been formed, in the completion of which the lover of fine arts, the student of natural history, the experimental philosopher, the mechanician, and antiquarian will alike find their taste gratified; where the old will find much that will speak to them of days long gone by, and at the same time remind them of the astonishing progress made in modern times; and where the young will find much to amuse and instruct, and at the same time to encourage and cheer them on in their path, by showing them in the results of art and science what talent and perseverance can produce, when directed to a good and noble end. The prices have been fixed so low as to enable both rich and poor alike to enjoy it.[25]

Here was a broad constituency, matched by predictions of what might be gained from a visit: affirmation of good taste, the production of knowledge, the generation of nostalgia, entertainment and optimism for the future. From a financial point of view it certainly made sense for the organisers to cultivate a wide base of support in order to liquidate the debts of the Mechanics' Institution. It had to be all things to all people and still be specific enough in its presentation of objects to attract those with narrower interests.

The *Leeds Times* was similarly forgiving of the unfinished state of the displays when it visited on 26 July 1843 (the original date on which it had been planned to open to the public) and reassured its readers that 'we are glad to say that, though then in a very incomplete state, as far as regarded the arrangement of the multitudinous object of interest which have been contributed, yet it was evident that the exhibition, when completed, will be one of surpassing variety, extent and interest'.[26] Any concerns held by the committee that there would be too few contributions were dispelled, to the extent that the *Leeds Times* worried that the organisers 'will rather feel themselves bewildered amidst their abundance of stores which they will find it difficult, with all their laborious exertions, to dispose and arrange in convenient order, in time for the opening of the exhibition'.[27] Its visit in the last frantic days of installation presented 'a busy scene of industry ... a large number

25 'Leeds Second Public Exhibition', p. 5.

26 'Leeds Public Exhibition', *Leeds Times*, 29 July 1843, p. 8.

27 'Leeds Public Exhibition', p. 8.

of operatives were at work, with hammers, ropes, ladders, pulleys, and all the appliances of mechanical labour, and indeed the whole place, in these respects, looked like some busy manufacturing establishment'.[28] This scene of real labour would be replaced by a gallery dedicated to the performance of labour once the exhibition was installed.

The Museum of Antiquities

With the exhibition now open to the public, we can attend to its content and arrangement. The first space entered by the visitor was the Ante-Room, installed as a Museum of Antiquities, or what the *Leeds Mercury* described as an 'Old Curiosity Shop', a reference to Charles Dickens's serialised novel published between 1840 and 1841.[29] The *Mercury* suggested that 'the visiter [*sic*] will at once be struck by the taste, accurate knowledge, and almost incredible industry through which this most rare and valuable collection has been formed and arranged'.[30] The newspaper also noted that it had been the committee's intention to cast the room as a baronial hall, 'but so great was the variety and number of objects sent, that it had to be changed'.[31] The shift in intent was a strategy to manage the interpretation of an otherwise miscellaneous body of objects by classifying them as 'curiosities'. Unlike the display and interpretation of curiosities in 1839, however, in 1843 an additional layer of meaning was constructed around these objects through the frame of antiquarianism. The fundamental organising principle was an idea of 'the past', reinforced through careful staging of objects, paintings and furniture, thought notable for their age, rarity and ability to operate together to create an impression of an undifferentiated bygone time. The *Leeds Times* interpreted the result as highly resonant and evocative, embedded with historicity:

> The time worn and richly storied relics of the past, have power to awaken noble and varied associations. The yellow and shrivelled parchment culled from ancient Archives—the rusty firelock picked from some field of famous battle—the careless scrawl, unsightly though it be, which is the impress of the mighty dead—these and such as these, are to be treasured, as spells of magic potency on

[28] 'Leeds Public Exhibition', p. 8.
[29] 'Leeds Second Public Exhibition', *Leeds Mercury*, 29 July 1843, p. 5.
[30] 'Leeds Second Public Exhibition', p. 5.
[31] 'Leeds Second Public Exhibition', p. 5.

> the creative spirit, which enable it to conjure up the forms and shadows of things bye gone [*sic*], to give substance and reality to its imaginings, and to live for a time in the world of its ancestors.[32]

These objects were invested with an almost supernatural ability to transport the viewer through time and materialise the past. They were presented as tangible witnesses to history, undeniable and irrefutable in their capacity to serve as documentary evidence.

The *Leeds Mercury* described the room in more prosaic terms:

> The upper part of the time-decayed walls is studded with armour of the time of the Commonwealth, interspersed with stag's heads, and rare and valuable portraits of the long departed; devices of arms, &c. &c. are grouped with taste; furniture of every kind, from the best time of the Tudors—(rare relics)—down to William and Anne, is everywhere placed with much effect.[33]

The impression reads as though the collection were the kind of accretion of different historical periods that might be found in a country house of a certain age. The atmosphere of the room was also determined by its lighting, described as 'subdued and sombre' and 'peculiarly appropriate'.[34] This use of lighting had two functions: to reinforce the impression of the space as an authentic reconstruction of the past—before it was possible to light a room with anything but candlelight—and also to shade the scene from too much scrutiny, creating an evenness of tone and message that would not so easily be disrupted by objects that might be considered out of place or not of the highest quality. Bude, Rose and Pelletan lights were used to keep the exhibition open until 10 p.m. each night.[35] The latter of these inventions had only been patented in May 1843, 'for the combustion of the mixed vapour of turpentine and water', which demonstrated that innovative technologies of display were also part of the message.[36]

Displayed around the walls of the Cabinet, portraits were an important part of the purpose of the exhibition because they represented exemplars of religious, political and scientific activity that were thought to provide

32 'An Afternoon in the Exhibition', *Leeds Times*, 19 August 1843, p. 6.

33 'Leeds Second Public Exhibition', *Leeds Mercury*, 29 July 1843, p. 5.

34 'Leeds Second Public Exhibition', p. 5.

35 'Leeds Second Public Exhibition', p. 5.

36 'The Pelletan Light', in *The Year-Book of Facts in Science and Art* (London: David Bogue, 1845), pp. 78–79.

ideals for the visitor to emulate. The *Leeds Mercury* highlighted a number of these portraits for special attention and noted the way in which the viewer's interests and beliefs might inform their response: 'The lover of science will dwell with interest upon a portrait of Dr. Priestley. The admirer of active benevolence and high public spirit will feel no less interest in a full length portrait of John Harrison, Esq., the founder of St. John's Church, in this town.'[37] It is significant that fifty years later, when Leeds was granted the status of a city in 1893, these two men were chosen to be the subject of bronze sculptures by Alfred Drury (1856–1944) and Henry Charles Fehr (1867–1940) respectively for the new City Square sculptural scheme instigated by Thomas Walter Harding (1843–1927), to which we will return in Chapter 7. The Third Polytechnic Exhibition of 1845 would introduce the subject of the third of the four bronze statues in the form of a marble bust of Walter Farquhar Hook by William Day Keyworth (c. 1818–97), who would be rendered in bronze by Frederick Pomeroy in 1902. We will come back to Keyworth's bust of Hook when we turn our attention to the 1845 exhibition in Chapter 3.

The Picture Gallery

The second space entered by the visitor was the Picture Gallery, described as 'radiant and almost dazzling with richness of beauty, containing a display of oil paintings, many of them by the old masters, such as is but rarely seen out of the metropolis'.[38] Although the term 'polytechnic' was applied less often to this exhibition, it did not mean that the display of fine art was completely separated from the other departments. Cases and tables along the sides of the Picture Gallery displayed geological specimens, shells, coins, small sculptures, glass and enamel and 'curiosities from China and India, New Zealand, &c.'[39] The exhibition was presented as an efficient and instructive surrogate for travel. There was no need to journey to London and process around country houses to see the finest works of art: here they were in Leeds, conveniently collated to allow comparison between artists, periods, styles and individual objects which would otherwise not be possible.

The *Leeds Times* was more moderate in its assessment of the relative merits of the Picture Gallery. It reported that the walls were 'completely covered with splendid pictures. They are more numerous that in the last

37 'Leeds Public Exhibition', *Leeds Mercury*, 19 August 1843, p. 5.

38 'Leeds Second Public Exhibition', *Leeds Mercury*, 29 July 1843, p. 5.

39 'Leeds Second Public Exhibition', p. 5.

exhibition, which, however, we think, contained one or two pictures that are unequalled by any in the present one.'[40] The *Leeds Times* highlighted works 'said to be by' Correggio, van Dyck, Bruegel, and Rottenhammer and Charles West Cope of Leeds.[41] Three weeks later the same periodical published a much more extensive and enthusiastic review of the Picture Gallery.[42] It drew attention to a large painting that was both first in the catalogue and first to be encountered at the entrance to the gallery: *Edwin, from Dr. Beattie's Minstrel* by Joseph Wright of Derby (1734–97).[43] The author described the painting and its merits as follows:

> The figure is of a youth, in rustic attire, but of gentle aspect, seated with his head leaning upon his hand, in thoughtful mood, but with an eye kindling with poetical enthusiasm. The picture is very beautiful, and highly finished. The colouring is somewhat subdued; the painter seems to have been perfectly successful in working out his conceptions; and has contrived to throw a pervading sentiment of poetical refinement over the picture.[44]

The term 'highly finished' occurred frequently in the critical reviews of the paintings and it was used to denote a convincing resemblance to nature in form and technique, related to a detailed and accurate surface and controlled application of paint. Commentators were pleased to see the stylistic particularities of different artists, as long as the traces of their hand and their labour remained concealed.

The ways in which Wright's work were praised stand in contrast to the terms used to lament the perceived shortcomings of another artist's work. John Rhodes (1809–42) was a local artist who had died on 3 December 1842, not long before the exhibition was conceived. Rhodes enjoyed some success in the 1820s and 1830s and exhibited at the Northern Society for the Encouragement of the Fine Arts in Leeds and the Royal Academy of Arts and the British Institution in London.[45]

40 'Leeds Public Exhibition', *Leeds Times*, 29 July 1843, p. 8.
41 'Leeds Public Exhibition', p. 8.
42 'An Afternoon in the Exhibition', *Leeds Times*, 19 August 1843, p. 6.
43 'An Afternoon in the Exhibition', p. 6. It is referred to as 'Beatties Minstrel by moonlight', in the article.
44 'An Afternoon in the Exhibition', p. 6.
45 Corinne Miller, 'Rhodes, John Nicholas (1809–1842), artist', *Oxford Dictionary of National Biography* (Oxford University Press, 2004), https://www.oxforddnb.com/view/10.1093/ref:odnb/9780198614128.001.0001/odnb-9780198614128-e-23452 (accessed 23 June 2020).

The critic at the *Leeds Times* wrote that some of his pictures on display 'possess great merit',

> but they are all strongly marked with his peculiar and somewhat unnatural style. They look rough and unfinished, as if dashed off at the moment he conceived the design, and never afterwards touched. Here is one of his best, "Sit up Sirrah!" a pleasant subject, and well conceived—a cottage door, with trees and sunshine, and rustic figures, and—what gives the name to the picture, a little roguish-looking country urchin making a spaniel dog sit up on its hind legs, with a pipe in its mouth, which it does with all becoming gravity and resignation, but with a somewhat wistful look, nevertheless, as if it would prefer being in a more natural, if less dignified position. The boy and the dog are capitally painted, and have a pleasant natural expression about them, but, unless the picture is viewed at some distance, it looks very unfinished.[46]

Had Rhodes been working as little as half a century later, he might have found the same work praised as expressive, perhaps even moving in the direction of early impressionism. In his own time Rhodes was judged to have been a competent painter, liked well enough for local patronage, with a career cut short by his early death. If Wright's work was celebrated for being 'highly finished' and Rhodes denigrated for appearing 'unfinished', there had to have been a recognisable point at which a painting was simply 'finished'. It was not enough for a painting to resolve itself at a distance; it had to be accurate, detailed and legible on close inspection too. The author of the article was not quite finished with the assessment of Rhodes and concluded with what they considered the best (or more accurately, the least worst) of his work on display:

> It represents a peasant boy going to the hayfield, with a basket on one arm and a black bottle swinging in the opposite hand. The figure of the boy is very natural and unsophisticated. He jogs along the pathway at an easy pace, with his head hanging on one side, happily undisturbed by anything like thought or care. So lazily serene he looks, that we doubt even if a stone on the pathway would tempt him into the momentary energy of giving it a kick forward—and that is a great temptation to a peasant lad. The straggling boughs of the blackberry, which stretch over the rude

46 'An Afternoon in the Exhibition', p. 6.

> rustic wall, and the broad leaved docks and rough grasses that grow at the bottom, give a good idea of a rural pathway. But the strange thing about the picture, is the almost ludicrous preponderance of blue, particularly in the unnatural colour of the sky. The artist, one might suppose, if such a thing were possible, had been labouring under a sort of blue jaundice, when he painted it, and unconsciously saw nature through that discoloured medium.[47]

The first two thirds of the paragraph are devoted to a descriptive account of the picture, which tend towards an imagined rural narrative on the part of the author. It is only the final sentence that discusses the formal qualities of the painting rather than its subject matter. There was no question in the mind of the critic that the artist might have intended to foreground the blue spectrum as a deliberate device. If the fundamental framework through which the merits of a work of art could be judged was truth to nature, any deviation resulted in the picture being read as false.

The article in which these critiques were published also entered into a wider discussion of the quality and status of British art, or more specifically English art. The author noted the attention afforded to two paintings by the Flemish Baroque artist Frans Snyders (1579–1657) as a result of their impressive scale and dominant position in the hang. While the article recognised the quality of the work, these hunting scenes were considered too graphic. It was an opportunity to contrast snarling, bloody scenes with the work of the British animal painter Edwin Landseer (1802–73): 'That great living artist is superior to them all. There is an individuality, a character about his animals, that you find in none else; and how exquisitely are they finished!'[48] Again the quality of finish was a primary consideration in the appraisal of a work of art. Here it clearly related to the extent to which the artist achieved a sense of realism in their work, as the author described having recently seen Landseer's paintings at an exhibition in Derby, where they 'were particularly struck with portraits of two favourite dogs belonging to the Duke of Devonshire. One was a spaniel, and so soft and silky, and real looked the hair, that it was hardly possible to resist the temptation to stroke it.'[49] The writer used this opportunity to lament the 'defective' lack of representation of work by living English artists, as their wider inclusion would have presented the opportunity 'to show the progress

[47] 'An Afternoon in the Exhibition', p. 6.
[48] 'An Afternoon in the Exhibition', p. 6.
[49] 'An Afternoon in the Exhibition', p. 6.

of art in our own country'.[50] A painting by John Martin (1789–1854) was presented as an exception that proved the rule. His work *Sadak in Search of the Waters of Oblivion* (1812) marked him, in the eyes of the writer, as 'an artist of sublime and original genius, of whom our country may well be proud'.[51] National pride was exactly what was at stake here, tempered by the nagging doubt that continental Europe remained firmly ahead in the art of both the past and the present. The author talked up Martin's painting in the following terms:

> What an awful scene has he conjured up in this small picture! Rocks heaped on rocks, towering to the heavens, shutting out the external world with vast walls of eternal granite. The waters gathered on the level at their base, sweep over the edge in two whelming cataracts, one of them, betwixt two rifted crags, into a lower deep. And here, in this "tremendous solitude," is "Sadak," lonely and desolate, clambering up a shelf of rock, betwixt the two cataracts, with wild and furious eagerness, as if a troop of fiends were in pursuit of him. The smallness of the figure gives one an idea of the vastness of the scenery, and as we contemplate it, a kind of awe creeps over the mind, which none have the same power to produce as John Martin—a power which, spite of all that his detractors may say, affords indubitable evidence of high and sterling genius.[52]

The dimensions of the picture—76.2 by 63.5 cm—made it not the smallest of paintings, but neither did it demand attention by virtue of size alone. Scale is implied by the tiny size of the figure, set against a vertiginous landscape.[53]

An artist who occupied a liminal position in the debates around the quality and significance of both historic and contemporary British painting was Anthony van Dyck (1599–1641), whose double portrait *Lord Strafford and His Secretary Sir P. Mainwaring* was displayed in the Picture Gallery. Described as 'a remarkable and powerful painting, and worthy of the great master by whom it was executed', the article in which this assessment was published went on to articulate further reasons why the visitor should pay particular attention to this particular picture:

50 'An Afternoon in the Exhibition', p. 6.
51 'An Afternoon in the Exhibition', p. 6.
52 'An Afternoon in the Exhibition', p. 6.
53 It has been in the collection of Southampton City Art Gallery since 1949.

> It possessed also great historical interest, for who does not wish to know the outward presentiment of the able, bad, but unfortunate Earl of Stafford [*sic*], whose career forms so interesting and important a portion of English history? And he is brought before us on the canvas here, with such an appearance of life and reality, and so dark and sombre he looks, that it would require no great stretch of imagination, in the dim twilight of some old hall, to fancy him, like the portrait in the Castle of Otranto, walking down from his frame to fright poor mortals. The portrait gives you the idea of a proud man, of grave and powerful intellect. On his ample brow, thought seems to lour, dark and dangerous as a thunder-cloud. He is pondering on matters of gravest import, while his "Secretary," with upward glance, is listening for the words he is about to prompt, in answer to the document he holds in his hand. Mark the success with which the painter has given the quick look of eager attention in the secretary, and the grave, fixed aspect of profound deliberative thought in the master. Though the secretary is a quick man, you see at once, and as if by a sort of contrast, the far more capacious intellect of Strafford. The picture is one of great interest and value.[54]

As we have already seen in the reception of the contents of the Museum of Curiosities through which the visitor first encountered the exhibition, part of the value of the painting was located in its status as a historical object. The writer assumed that the appearance of the subjects in the painting—Thomas Wentworth, 1st Earl of Strafford (1593–1641) and Sir Philip Mainwaring (1589–1661)—accurately resembled their appearance in life. This might be considered a reasonable assumption on the basis that van Dyck was widely celebrated as a portrait painter at the highest level during his lifetime. As photography remained in its infancy at the time of the 1843 exhibition, painted portraits and sculpted portrait busts and statues were part of a limited range of media through which to record the appearance of a person for posterity. As such they were considered as much part of the historical record for their documentary role, as they were works of art embedded with layers of mediation between sitter, artist, materials and the social, economic and political context of production. What the writer alluded to was that this double portrait provided a point of access to allow the viewer to visit an imagined past and to speculate on the relationship between

[54] 'An Afternoon in the Exhibition', p. 6.

the appearance of the two men and their character, based on prior knowledge of the trajectory of their lives.

The Cabinet

The display of pictures continued in the Cabinet, a smaller space containing more oil paintings and watercolours, including those described as having belonged to 'Turner's early days, unequalled in truth, power, and brilliancy, by any other in the style of the arts'.[55] As had been the case at the 1839 exhibition, Turner's reputation remained high. The *Leeds Times* drew attention to one picture in the Cabinet for special consideration: *Venus and Adonis*, loaned by Mr Henwood.[56] An article about the painting had been published by the newspaper the previous month, under the title 'Singular History of a Fine Painting'.[57] The article described how Henwood, a 'restorative artist', came to own the painting: 'The picture along with forty or fifty others was received from the Continent, as security for a debt by a respectable merchant in Leeds, and after remaining in his warehouse unheeded for some years, was ultimately sold by him to Mr. Henwood, for four pounds'—the equivalent of approximately twenty days' wages for a skilled tradesman at the time.[58] The state of the painting when Henwood purchased it was described as 'exceedingly dirty' and when he attempted to resell the picture at a small profit, 'so completely were its beauties concealed from observation, that Mr. Henwood ... was taunted with having given £3 17 6 too much for his bargain'.[59] Having failed to sell the picture for anywhere near as much as he paid for it, Henwood undertook a restoration of the surface, during which 'its beauties gradually dawned upon the artist', as patches of white paint gave way to 'a production of high finish and poetical conception'.[60] It was suggested that areas of white paint had been deliberately applied to obscure the quality and value of the painting, and Henwood speculated that the original owner had sought to protect it from Napoleon's acquisitive campaigns. Once the picture was restored Henwood sent it to London, 'where it was

55 'Leeds Second Public Exhibition', *Leeds Mercury*, 29 July 1843, p. 5.
56 'Leeds Public Exhibition', *Leeds Times*, 29 July 1843, p. 8.
57 'Singular History of a Fine Painting', *Leeds Times*, 10 June 1843, p. 5.
58 'Singular History of a Fine Painting', p. 5. Purchasing power calculated using the National Archives Currency Converter: 1270–2017, https://www.nationalarchives.gov.uk/currency-converter/.
59 'Singular History of a Fine Painting', p. 5.
60 'Singular History of a Fine Painting', p. 5.

immediately pronounced by several leading artists and academicians to be an original production of Albano, an eminent Italian artist'.[61] Francesco Albani (or Albano) (1578–1660) was a Bolognese Baroque painter who produced, along with his wider circle of followers, a number of iterations of the *Venus and Adonis* theme. Following its attribution, given weight—although not evidence—through a vague association with the Royal Academy of Arts and unnamed 'leading artists', the article reported that Henwood had received a generous proposal: 'One hundred guineas have been offered for it by a gentleman of this town, and refused by the owner.'[62] The piece concluded with the information that the work was likely to be displayed at the exhibition the following month. It is difficult not to conclude that the underlying intention of the article was to talk up the painting to sell at the exhibition or after its display, presumably for a sum greater than 100 guineas.

Once the painting had been on display for almost a month the *Leeds Times* published another extended notice. The reporter greeted the picture as an old friend, 'with a quick gladness, such as the renewal of a pleasant and well-remembered acquaintance inspires', and described it as 'an exquisite painting, to which we have, ere now, in these columns, paid our tribute of admiration'.[63] The writer went on to provide a long description of the painting, with particular attention to the figure of Venus. It concluded that 'the whole picture is beautifully finished' and that it was 'the production of Albano'.[64]

Henwood sought to capitalise on the exhibition in another way too; by advertising his services through three other paintings he had restored that were also on display. An advertisement appeared in the *Leeds Mercury* towards the end of the exhibition with the following copy:

> G. Henwood, Restorative Artist, 11, Basinghall Street, Leeds, begs to call the Attention of the Nobility and Gentry to his peculiar Method of restoring old paintings, which has met the Approbation of the late Davies Gilbert, Esq. President of the Royal Society, and other Connoisseurs; and he solicits Attention to the following Specimens in the Leeds Public Exhibition—Nos. 5 and 105 in the Gallery, and Nos. 46 and 61 in the Cabinet, all of which have been retrieved from states in which their Proprietors had deemed them worthless, and

[61] 'Singular History of a Fine Painting', p. 5.
[62] 'Singular History of a Fine Painting', p. 5.
[63] 'An Afternoon in the Exhibition', *Leeds Times*, 26 August 1843, p. 6.
[64] 'An Afternoon in the Exhibition', p. 6.

> who have kindly offered References. Carving and Gilding of every Description on the lowest possible Terms compatible with good Workmanship.[65]

The advertisement used some of the same strategies as the article which first drew attention to the *Venus and Adonis* in his possession: recourse to expertise and connoisseurship located in the capital alongside examples of paintings considered to be without worth before his restoration of them. It is possible that Henwood's association with the exhibition informed his decision to expand his business. Two and a half years later, in April 1846, the *Leeds Times* announced that Henwood had opened his own commercial gallery:

> Leeds is by no means eminent in the above department [fine art];—in fact we stand in the fine arts at a sad discount. An opportunity, however, is now offered, which if carried on with spirit, and duly performed, will be calculated to redeem, in some measure, our evident defect. A permanent gallery for the exhibition and disposal of pictures, is just opened, by our talented friend, Mr. Henwood. We have been favoured with a slight view of some of the pictures, and we can recommend them to the connoisseur as truly worthy of attention… We can fairly promise our readers a treat, and the connoisseur an opportunity, rarely afforded, for the study of various styles and subjects.[66]

The outcome of this venture appears not to have been successful, as George Thomas Henwood, then described as a 'carver and gilder, dealer and chapman', was declared bankrupt on 22 July 1847.[67]

Aside from *Venus and Adonis*, another painting displayed in the Cabinet provided a further opportunity to discuss the question of attribution and authenticity. The *Leeds Times* noted a number of other pictures as worthy of attention in this room, including *Madonna* by Guido Reni (1575–1642). The work was described as 'a lovely painting', although the reporter noted that some thought it to be a copy. In any case they continued, 'we care little which it is; there can be no doubt of its being a beautiful and highly finished painting, and one, which, Guido, were he living, might well be proud to acknowledge'.[68] The point

65 Advertisements and Notices, *Leeds Mercury*, 9 September 1843, p. 1.

66 'Fine Arts', *Leeds Times*, 4 April 1846, p. 4.

67 'George Henwood', *Jurist*, vol. 11, part 2 (London: S. Sweet, 1848), p. 304.

68 'An Afternoon in the Exhibition', p. 6.

was not whether the painting were an original or a copy; authenticity was not what was at stake. What the work communicated through its subject matter, composition and technique were what mattered in this context. The same article went on to discuss the distinction between originals and copies and confirmed that what counted in this context was the quality of the copy rather than its status and authorship. The reporter drew attention to a number of paintings recorded as being by Giovanni Antonio Canal, known as Canaletto (1697–1768).[69] The journalist entered the Saloon and took notice of four paintings called *Views in Venice* purported to be by Canaletto. Here follows the stinging assessment of their quality as works of art:

> If the Adriatic were a pool of stiff blue mud, and the sky of Venice were made of sheet lead, then might these be views of Venice, and painted by Canaletti [*sic*]. But compare them for a moment with another view in Venice, by Canaletti, in the Picture Gallery, No. 100, and mark the glittering transparency of the water, the clearness of the sky, the fine perspective of the buildings, and the air of reality thrown over the whole, in the latter picture, as compared with the leaden dulness of those in the saloon; and, though no connoisseur, you will without difficulty pronounce correctly betwixt the genuine and spurious Canalettis.[70]

The author appears to suggest that the question of attribution was not the primary concern, but rather that the *Madonna* attributed to Guido was convincing, conformed to a certain standard of technical proficiency and communicated its message to the viewer. By contrast, the contested works attributed to Canaletto were criticised not because they were likely to be copies, but because they were bad paintings. That is to say, they were not judged to sufficiently resemble the style and technical facility of their purported maker and therefore failed to communicate the idea of Canaletto's Venice that was so distinctive and celebrated.

The Cabinet provided the setting for the first work by a woman on display in the exhibition to be noted in the periodical press, albeit only briefly: 'There are some pleasing and well executed water colour drawings here by Mrs. Anson, a lady, who resides in this town.'[71] It is

[69] Katherine Baetjer, '"Canaletti Painting": On Turner, Canaletto, and Venice', *Metropolitan Museum Journal*, 42 (2007), 163–72 (p. 163).

[70] 'An Afternoon in the Exhibition', p. 6.

[71] 'An Afternoon in the Exhibition', p. 6.

possible that they were considered noteworthy by virtue of the artist being a woman, but not noteworthy enough to be given a comprehensive description or analysis. In this respect it is more likely that they occupied the territory of the novelty or curiosity; exceptional in the most prosaic meaning of the term. The *Leeds Mercury* provided a little more information when it noted that seven watercolours by Mrs Anson, 'a lady lately come to reside in Leeds', were on display in the Cabinet. There was also an important strategic reason for their inclusion: 'They display a high degree of talent, a fine perception of the natural and the true, and are all drawn with great nicety and accuracy. They are on sale, and the talented authoress, we understand, is desirous of giving instructions in this noble art.'[72]

By contrast, a painting by Thomas Webster (1800–86) titled *The Grandmother* received a great deal more detailed attention.[73] Dated 1842, it was one of the most contemporary paintings on display, but far from engaging in the present moment, *The Grandmother* presented a character study of the sort that has come to represent Victorian art at its most sentimental. In a domestic setting the eponymous grandmother is teaching her young grandson to read, but the viewer can see what the grandmother cannot: that the boy's eyes are not following the words on the page as directed by his grandmother's knitting needle, but rather meet the gaze of their audience with an expression that appears to implore them to rescue him from his studies. The critic of the *Leeds Times* saw beyond the simple narrative and celebrated the artist's technique, capacity to communicate feeling and the painting's native qualities:

> The picture is exquisitely finished. The boy seems almost real, and everybody who looks at the picture instantly recognises and enjoys the truthfulness of the expression. It is a truly English picture, and such as English people, above all others, can enjoy. It is perfect of its kind.[74]

Once again the quality of finish and truth to life were singled out for particular praise. For the critic, the artist had achieved both a sense of reality through the figures, objects and setting of the painting and,

72 'The Leeds Public Exhibition', *Leeds Mercury*, 2 September 1843, p. 5.

73 The painting was sold under the title *Grandma Knows Best* by Sworders as Lot 274 in 'The Winter Country House Sale' at Stansted Mountfitchet on 9 December 2014.

74 'An Afternoon in the Exhibition', p. 6.

on the other hand, a truthfulness in the relationship, situation and behaviour. From this perspective the picture was faithful in form and message, although not necessarily universal in its appeal. To assign it an 'English picture' most enjoyed by an English audience suggests that at least part of its appeal was located in its familiarity. It relied on the tropes of the recalcitrant student, the domestic setting and the kindly—if strict and exacting—figure of the grandmother. That it was considered to be perfect 'of its kind' provided an important qualifier; this was not a work of art that aimed to communicate beyond its local social, economic or political boundaries.

The unnamed author who wrote about the exhibition for the *Leeds Times* did not operate in a professional capacity as a critic. The development of judgement and taste were central to the broader purpose of the exhibition, and having stated that their conclusions were informed by personal preference rather than any sort of external measure, they argued that the imposition of such measures were detrimental to audiences and to English art more broadly:

> In our remarks we have been guided by no great technical knowledge of the art, but have simply given the impressions produced upon our own minds by the pictures themselves, and though some of the opinions may be deemed erroneous, they have at least the merit of being such as were really felt at the time they were expressed. There is not a little truth in the homely observation, that "that is the best picture, which pleases best." There can be no doubt, at all events, that it is best to the individual himself whom it pleases best; and we would much rather hear a man express his own genuine opinion, even at the risk of incurring censure, then [*sic*] launch out into indiscriminate eulogies of some painting which he has heard is very fine, but whose excellencies he really does not understand ... though he may go on increasing in knowledge, and in taste, let him ever express the genuine sentiments of the moment, otherwise his opinions are of no intrinsic value whatever, for they are mere echoes of the opinions of others. In this way alone, is a good national, as well as individual taste to be formed; and it is our conviction that the English school of painting has been much injured, both in originality, and every other excellence, by the slavish servility with which painters and critics have subjected their own judgements to certain professional standards of taste, and canons of criticism. There is a pleasure in hearing even the rough natural exclamation of the rustic or mechanic, unsophisticated by prejudice, as he views the paintings

> in the exhibition; because we know that it is a genuine expression of opinion and taste, and as such, it has value, though it emanates from an uncultivated intellect. It shows that the picture has that in it, which is calculated to produce an effect; it must have something good about it.[75]

According to the author, the response to a work of art was generated from two directions: from the inherent qualities of the art object and from the individual's encounter with it. The argument that the validity of judgement should be based on sincerity of expression rather than being a product of education and conformity operated in contrast to the emerging design reform movement that sought to standardise public taste through exposure to objects considered to embody and communicate 'correct principles'.[76]

The sentiments expressed in the article point towards an underlying antipathy towards the centralisation and prescription of matters of taste and judgement. According to Woodson-Boulton, the opposing Ruskinian approach took 'an understanding of art as essentially about the uncovering of universal truths through the search for ideal beauty, both being based on unchanging natural law. While the artist alone could reveal the essential harmonies of nature, all of humankind could recognize them.'[77] The *Leeds Times* was not alone in its promotion of individual judgement over expertise. The *Leeds Mercury* similarly distanced itself from a position of superior knowledge in its notices: 'We disclaim all familiarity with what are termed artistical effects, and we shall therefore only speak of such points of the pictures as would strike an ordinary observer.'[78] It did, however, recommend that visitors be 'accompanied by that useful guide, the exhibition catalogue'.[79]

75 'An Afternoon in the Exhibition', p. 6.

76 See Suga Yasuko, 'Designing the Morality of Consumption: "Chamber of Horrors at the Museum of Ornamental Art, 1852–53', *Design Issues*, 20:4 (2004), 43–56. Clive Wainwright suggested that the work of their contemporary and colleague Augustus Welby Northmore Pugin informed both the rhetoric and practice of the design reformers, specifically *The True Principles of Pointed or Christian Architecture*, published in 1841. See Clive Wainwright, 'Principles True and False: Pugin and the Foundation of the Museum of Manufactures', *The Burlington Magazine*, 136:1095 (1994), 357–64.

77 Amy Woodson-Boulton, *Transformative Beauty: Art Museums in Industrial Britain* (Palo Alto, CA: Stanford University Press, 2012), p. 8.

78 'The Leeds Public Exhibition', *Leeds Mercury*, 26 August 1843, p. 5.

79 'The Leeds Public Exhibition', p. 5.

This suggestion was also motivated by the fact that the catalogue was published by William Baines, a relation of Edward Baines, editor of the *Leeds Mercury* between 1801 and 1848, and his son, Edward Baines Junior, who was deeply involved in the voluntary societies of Leeds (including his vice-presidency of the Leeds Mechanics' Institution and Literary Society) and their exhibitions.[80]

The Saloon

On leaving the Cabinet the visitor entered the largest space in the exhibition: the Saloon, also known as the Concert Room. As in the 1839 exhibition, the Saloon operated as a gallery with a number of functions and displayed the greatest diversity of objects presented together in the same space. Here the exhibition was in motion, with a tank filled with water to show model ships and steamboats encircled by a track with working models of steam engines and locomotives travelling across a miniature railway. Specimens of natural history, philosophical instruments and minerals were displayed underneath the gallery that provided views from above down into the Saloon, alongside a confectioner selling treats to visitors. The confectioner's stall was illuminated by a 'brilliant gas star, the gas for which is gratuitously furnished by the Old Gas Company'.[81] What this seemingly incongruous pair of exhibits shared was the processing of a raw material into a commodity; it was not just a season ticket and a catalogue that visitors were encouraged to purchase, and consumption was not limited to a visual encounter. In the section of the Saloon usually reserved for the orchestra was the organ brought in especially for the exhibition and its concert programme. Most intriguing of all was the prospect that 'in this room has likewise taken up her abode *an invisible young lady*, who, we feel persuaded, will exercise her various and extraordinary powers upon many a visiter [*sic*]'.[82] The gallery above the Saloon included an electrical machine, further specimens of natural history and, most impressively, a painted scene of the Bay of Naples along with a working model of Mount Vesuvius which erupted periodically.[83]

80 David Thornton, 'Mr Mercury—A Biographical Study of Edward Baines with special reference to his role as Editor, Author and Politician' (unpublished doctoral thesis, University of Leeds, 1999), p. iii.

81 'Leeds Second Public Exhibition', *Leeds Mercury*, 29 July 1843, p. 5.

82 'Leeds Second Public Exhibition', p. 5.

83 'Leeds Second Public Exhibition', p. 5.

The Saloon also contained a number of works of fine art. One of the objects most remarked upon was a sculpture of a boy, carved in stone at near to life size by Alfred Bromley (c. 1828–87). The subject was the brother of the sculptor and it was the sculptor who attracted as much, if not more, attention than the work by virtue of his age and relatively low professional status: he had carved the work as a 13-year-old apprentice at the firm of Welsh and Lee. The *Leeds Times* wrote, 'though somewhat rough, it is a surprising production for a boy of his age'.[84] He exhibited again at the 1845 exhibition, and the *Leeds Intelligencer* positively reviewed the new work, while reminding readers of the sculpture displayed two years earlier:

> During the past week, several valuable additions have been made to the exhibition, amongst which we particularly notice the marble bust of a gentleman of Leeds, cut by Alfred Bromley, the youth who executed the statue of his brother in stone, which attracted such universal attention at the last exhibition. We understand he is an apprentice to Messrs. Welsh and Lee, marble masons, of this town, and it gives us great pleasure to award him that meed of praise which his industry and talent so richly deserve, and we are of opinion that if his studies were directed in the proper channel, by some experienced hand, he would in all probability at some future period rise to considerable eminence in the profession for which it appears nature has peculiarly fitted him.[85]

Bromley did go on to fulfil his early promise as a sculptor. Born in Birstall, approximately seven miles south-west of Leeds, Bromley completed his apprenticeship with Welsh and Lee and went into business under his own name by the time he was 23.[86] Anthony Welsh (1823–91) and Dennis Lee (c. 1816–73) were prominent stonemasons in Leeds, manufacturing statuary and funerary monuments until their partnership was dissolved in 1870.[87] Bromley graduated from this profession into the production

[84] 'Leeds Public Exhibition', *Leeds Times*, 29 July 1843, p. 8.

[85] 'The Exhibition', *Leeds Intelligencer*, 23 August 1845, p. 5.

[86] 'Alfred Bromley', *Mapping the Practice and Profession of Sculpture in Britain and Ireland 1851–1951*, University of Glasgow History of Art and HATII, online database 2011, http://sculpture.gla.ac.uk/view/person.php?id=msib5_1213712634 (accessed 21 June 2020).

[87] 'Lee & Welsh', *Mapping the Practice and Profession of Sculpture in Britain and Ireland 1851–1951*, University of Glasgow History of Art and HATII, online database 2011, http://sculpture.gla.ac.uk/view/organization.php?id=msib5_1213898320 (accessed 21 June 2020).

of portrait busts and public sculpture. In 1853 he exhibited at the short-lived Leeds Academy of Arts discussed in Chapter 4, and by 1861 he had moved to Chelsea in London and exhibited twice at the Royal Academy of Arts.[88]

'Machinery in active operation'

The Saloon provided access to the rooms in an adjoining building, pressed into service to assuage the concerns of lenders of fine art about having their pictures displayed alongside working machinery. The two rooms had been added at considerable cost, presumably with the intention of recouping the expense through more impressive loans and a corresponding increase in visitors. One of the rooms was dedicated to 'machinery in active operation' and the *Leeds Mercury* described the scene:

> This will be truly a scene of life and motion: the turner, the silk-weaver, the file-cutter, the jacquard loom weaver, the comb-maker, the sheet and fillet card maker, and the electrotyper, will vie with each other in exhibiting their various arts and handicrafts. The whole is set in motion by a beautiful steam engine of four horse power, with a vibrating cylinder, made for the exhibition by Messrs. Kitson and Co.[89]

It is significant that the term 'motion' is used twice in this short description, with its sense of the constant movement of the factory floor, scaled down and sanitised. Although living exhibits were included in 1839, this component had been significantly extended in 1843. This was perhaps a result of its popularity with visitors, in combination with lobbying from manufacturers keen not only to show off their machines and what they were capable of producing, but also how content, productive and skilled their workforce were. The second of the two additional spaces contained a number of exhibits that had been displayed in the Saloon in the 1839 exhibition, including a diving bell, oxy-hydrogen microscope and dissolving views. The diving bell had been substantially enlarged to accommodate up to four people, 'at the small additional charge of sixpence, to pry into the secrets of

[88] 'Alfred Bromley', *Mapping the Practice and Profession of Sculpture in Britain and Ireland 1851–1951*.

[89] 'Leeds Second Public Exhibition', *Leeds Mercury*, 29 July 1843, p. 5.

the "vasty deep"'.[90] Also manufactured by Kitson & Co., the cost of this experience was the same as a single ticket to the whole exhibition, which limited the number of visitors who could afford to participate. In addition to the concert programme the committee organised a series of live scientific experiments, all of which were designed to encourage repeat visits.

The *Leeds Mercury* concluded its initial assessment with the following remarks: 'Here will be found enough to charm the eye, instruct the mind, and improve the taste; and from the drudgery of daily toil the visiter [*sic*], whatever his rank or pursuit in life, will here find ample relaxation.'[91] Instruction and improvement were familiar intents for the Victorian exhibition, but relaxation was slightly different. The exhibition functioned to mitigate the 'drudgery of daily toil' that supported industrial capitalism. By constructing a space that separated the labourer from their labour—and celebrated the ingenuity and nobility of that labour and provided concrete demonstrations of how it connected to the arts and sciences—the working-class visitor was to be left refreshed by their experience and ready to begin a new working day with increased pride and productivity. More important still, the exhibition steered them away from thoughts of social, political and economic revolution, so close to the surface during the 'Hungry Forties' and the concomitant rise of Chartist activism, which the historian J.F.C. Harrison argued emerged in Leeds in a more organised way than in other comparable urban centres with the decision to establish a Leeds Working Men's Association in 1837.[92]

Additions and subtractions

In addition to the short delay between the date on which the committee announced that the exhibition would open to the public and the date that it actually opened, the local periodical press stressed that its early reports were published before it was considered complete. Whether the exhibition could ever have been considered finished is open to question; it was certainly never static. Objects continued to flow in and out of the display for the whole duration in a way that would be almost inconceivable now, with object checklists, loans and exhibition designs

90 'Leeds Second Public Exhibition', p. 5.

91 'Leeds Second Public Exhibition', p. 5.

92 J.F.C. Harrison, 'Chartism in Leeds', in *Chartist Studies*, ed. by Asa Briggs (London: Macmillan, 1959), pp. 65–98 (p. 65).

agreed and fixed far in advance. Rather than being perceived as rushed and disorganised, the exhibition's mutability was considered a point in its favour and an inducement for visitors to purchase season tickets to see the latest arrivals. The *Leeds Mercury* noted the first changes to the display one week after it had opened:

> This delightful place of resort was opened to the public on Saturday last, but the arrangements at that time were not quite complete, owing to the multiplicity of duties devolving on the gentlemen who have so liberally expended their time in getting up the Exhibition. For the first three or four days, therefore, it was not practicable to observe the same regularity in the different periodical displays of interesting objects as was desirable, but happily the arrangements are now complete, and visiters [*sic*] may depend on witnessing any particular objects which they are anxious to see, at the same time specified in the daily bills of fare. Yesterday a number of beautiful models of shipping arrived from the Admiralty house. Up to Thursday night (only the fifth day since the opening of the Exhibition) about 4,000 season tickets had been disposed of, which at 2s. 6d. each, would realise the sum of £500! In addition to this, up to the same period, £50 10s. 6d. had been taken in six-penny admissions.[93]

A number of significant changes were made to the exhibition in September 1843.[94] The nature of the alterations to the Picture Gallery were not specified, although the following week the *Leeds Mercury* reported: 'Several pictures of inferior merit have been taken down, and replaced by others of the highest character.'[95] These pictures were joined by specimens of papier-mâché in the Picture Gallery and the Cabinet, 'which almost deserve to rank amongst the pictures of a very high order. They are decidedly the most beautiful specimens of this manufacture we have ever seen, and show to what a high degree it has been carried.'[96] Additions were also made elsewhere: a locomotive manufactured by Robert Fourness of Leeds was brought into the Saloon, new musical attractions were added to the concert programme and Chinese fireworks were set off 'without smoke, smell, or gunpowder'.[97] On 25 September

93 'Leeds Public Exhibition', *Leeds Mercury*, 5 August 1843, p. 5.
94 'Leeds Public Exhibition', *Leeds Mercury*, 9 September 1843, p. 4.
95 'Leeds Second Public Exhibition', *Leeds Mercury*, 16 September 1843, p. 5.
96 'Leeds Second Public Exhibition', p. 5.
97 'Leeds Second Public Exhibition', p. 5.

the dissolving views were exchanged for a new set, which were said to 'far exceed the old ones in delicacy, variety, and beauty', alongside a panorama of the bombardment of Algiers. The panorama was likely to be—or to have been inspired by—Charles Marshall's *Grand Marine Peristrephic Panorama of the Bombardment of Algiers* (c. 1823), which had been widely exhibited in public since the 1820s.[98] An increase in the number of visitors was attributed to these new attractions and included notable visitors Baron de Nieuman, the Austrian Ambassador, and the war hero Major Waymouth, both of whom were escorted around the exhibition by William Beckett.[99]

In the first six days a total of 14,285 visits were made to the exhibition. The interactive elements of the display proved popular: in the first few days of operation, the diving bell transported 200 people underwater, '50 of that number being ladies'.[100] The popularity of the exhibition could also be a cause for concern; on the first Thursday two men were caught trying to pick a woman's pocket. The *Leeds Times* noted the distinctive appearance of the itinerant larcenists: 'They are travelling thieves; one of them wore a blouse, and the other had but one arm.'[101] The following month a reader wrote to the *Leeds Times* to complain about those 'who are in the practise of cutting the ladies' dresses, apparently with scissors of some other sharp instrument ... his daughter visited the Exhibition on Tuesday, and while there, her bag was cut right across the bottom—it is supposed with the view of extracting the contents'.[102] In contrast to these reports of petty theft, the only other report of criminal behaviour was the transfer of two season tickets from their owners to other parties for unauthorised entry, for which the tickets were confiscated.[103]

98 See David Worrall, *Harlequin Empire: Race, Ethnicity and the Drama of the Popular Enlightenment* (Abingdon: Routledge, 2015), p. 114; John Plunkett, 'Moving Panoramas c. 1800 to 1840: The Spaces of Nineteenth-Century Picture-Going', *Interdisciplinary Studies in the Long Nineteenth Century*, 19 (2013), 1–34 (pp. 17–18, 20–21, 23–25, 32–34); and Erkki Huhtamo, 'Penetrating the Peristrephic: An Unwritten Chapter in the History of the Panorama', *Early Popular Visual Culture*, 6:3 (2008), 219–38 (pp. 220–21).

99 'Leeds Second Public Exhibition', *Leeds Times*, 23 September 1843, p. 5.

100 'Leeds Public Exhibition', *Leeds Mercury*, 5 August 1843, p. 5.

101 'Caution', *Leeds Times*, 5 August 1843, p. 5.

102 To Readers and Correspondents, *Leeds Times*, 2 September 1843, p. 4.

103 'Leeds Public Exhibition', *Leeds Times*, 12 August 1843, p. 5.

The second season

The exhibition was due to close on Wednesday 25 October 1843.[104] Unlike the 1839 exhibition, it soon reopened for a second season that lasted for a further four weeks, with new season tickets issued at 2s. 6d.[105] The *Leeds Mercury* promised: 'Few articles of value have been recalled, and these will be more than replaced by new objects. In every department the committee are endeavouring to produce novelties and striking improvements.'[106] While the first season was intended for the financial benefit of the Mechanics' Institution, the profits of the second season were to be dedicated to different causes: the Dispensary and the House of Recovery. The committee invited representatives of the two institutions to act as curators: Henry Chorley and Arthur Lupton on behalf of the Dispensary and T.B. Pease and Beverley Hill on behalf of the House of Recovery.[107] The economic purpose of the exhibition subtly shifted from a cause that was well-supported but not necessarily essential, to a cause that was described as 'pure charity'.[108] Visitors were urged to attend on that basis, coupled with the prospect of new novelties and a fresh programme of concerts alongside a request for further donations of 'attractive objects of art or science' to augment the displays.[109] The additional effort was required because the revenue from the first season had not met expectations, although the *Leeds Mercury* was keen to emphasise the success of the exhibition in higher terms:

> The total receipts, though falling short of what the warm friends of the institution, for the use of which the net produce will be applied, could have wished, have been considerable, and will no doubt leave a handsome surplus. The pecuniary part of it is no

[104] Advertisements and Notices, *Leeds Mercury*, 21 October 1843, p. 1.

[105] 'Leeds Public Exhibition.—Second Season', *Leeds Mercury*, 21 October 1843, p. 5.

[106] 'Leeds Second Public Exhibition', *Leeds Mercury*, 28 October 1843, p. 5.

[107] 'Leeds Public Exhibition.—Second Season', p. 5. Chorley (1810–78) was a surgeon and magistrate best-known for his role in the 'Leeds Dripping Riot' when he pursued the prosecution of his cook, Eliza Stafford, on learning that she had sold dripping from his household; Lupton (1809–89) was from a prominent Leeds family, descended from landed gentry and emerging as merchants, manufacturers and politicians in the nineteenth century; Pease was a merchant connected with the emerging railway network; and Hill was an agent at the Leeds and Yorkshire Fire & Life Company, who died aged 69 in 1848.

[108] 'Leeds Public Exhibition.—Second Season', p. 5.

[109] 'Leeds Public Exhibition.—Second Season', p. 5.

> doubt an important consideration, but the far higher consideration of having provided, for a long period, a cheap and rational resort to thousands of people, having given the means for improvement in the arts, for emulation in the walks of science, has beyond all doubt been answered; and, in this will all those who have, with such noble and generous self-sacrifice devoted their time and talents, find their best and only desired reward.[110]

The last few days were well attended, not least because the committee had advertised a special rate of 2d. per Sunday School student.[111] On 20 October 1843 2,000 students visited as a result, with children from workhouse schools provided with free entry.[112]

Between the closure of the first season and the opening of the second, more space was made in the Saloon for evening concerts through the removal of the fountain and relocation of the canal.[113] With only two weeks left until the final closure, it was clear that attendance had again failed to match expectations. The *Leeds Times* issued the following plea to its readers:

> We are happy to learn that the receipts of the last few days, though falling far short of what either of the friends of the Charities could have wished, or what the unremitting exertions of the Managers deserve, are nevertheless much more satisfactory that at the commencement, and we again make our earnest appeal to the public to give liberally their patronage and support, where so much is done to gratify them, and where the funds are devoted to so noble a purpose. The attractions of next week fall in no way short of the preceding one. Every evening a high musical treat is afforded.[114]

In support of its appeal the newspaper reported on the installation of a new attraction: Armstrong's hydro-electric machine, 'similar to the one now exhibiting in the Polytechnic Institution, London, and which is the wonder and admiration of every beholder, from the most scientific to the entirely uninitiated'.[115] The article emphasised that

[110] 'Leeds Second Public Exhibition', *Leeds Mercury*, 28 October 1843, p. 5.
[111] 'Leeds Second Public Exhibition', *Leeds Mercury*, 14 October 1843, p. 5.
[112] 'Leeds Public Exhibition.—Second Season', p. 5.
[113] 'Leeds Public Exhibition.—Second Season', p. 5.
[114] 'Leeds Public Exhibition for the the House of Recovery and Dispensary', *Leeds Times*, 11 November 1843, p. 5.
[115] 'Leeds Public Exhibition for the the House of Recovery and Dispensary', p. 5.

the machine could not be seen anywhere else outside the capital, and further: 'The mass of electricity produced in this way, far exceeds anything ever produced by means of any other instrument, and the effects of it, as may be imagined, are astounding.'[116] William Armstrong (1810–1900), later 1st Baron Armstrong, was a Newcastle-born scientist and engineer who discovered that steam could carry a positive electric charge under certain conditions. His machine was well suited for public display because it produced dramatic sparks, although it found no direct practical application and was quickly made obsolete by new technologies based on electromagnetism.[117] The effects of the machine in operation were promoted in the *Leeds Mercury*, which encouraged visitors to witness: 'The brilliant effect of the electric spark, whilst illuminating different devices—setting fire to various bodies—charging with astonishing rapidity a battery of jars—and identifying its power with "Heaven's own lightning".'[118]

In the final days of the exhibition a further feature that sat between technology and theatrics was added: every evening 'Mons. D'Auvigney' would display 'unrivalled automaton figures'.[119] This was not the first time this spectacle had been seen in the town, as it was reported that 'many of our readers may probably recollect having seen [them] with pleasure, under the direction of M. Maffey'.[120] Mechanical theatres were popular attractions and the Maffey family had been engaged in their display in France since the late eighteenth century, performing across Europe and the United States into the nineteenth century.[121] The exhibition formally closed on the evening of Saturday 25 November 1843, but 'owing to the unprecedented success which the representations of Mons. D'Auvigney and Mr. Shaw have had, and the generally expressed disappointment at their very short stay, the Curators have been induced to continue this part of the Exhibition for five days longer'.[122] The *Leeds Mercury* provided

[116] 'Leeds Public Exhibition for the the House of Recovery and Dispensary', p. 5.

[117] Ian McNeil, ed., *An Encyclopedia of the History of Technology* (London: Routledge, 2002), p. 351.

[118] 'The Exhibition', *Leeds Mercury*, 18 November 1843, p. 5.

[119] 'The Exhibition', p. 5.

[120] 'The Exhibition', p. 5.

[121] Erikki Huhtamo, 'Mechanisms in the Mist: A Media Archaeological Excavation of the Mechanical Theatre', in *Media Archaeology and Intermedial Performance: Deep Time of the Theatre*, ed. by Nele Wynants (Cham: Palgrave Macmillan, 2019), pp. 23–82 (p. 60).

[122] 'Leeds Public Exhibition', *Leeds Mercury*, 25 November 1843, p. 5.

a robust endorsement of the experience of seeing D'Auvigney and Shaw's act and described the effects of a magical spectacle in a rational age:

> In witnessing these performances, every one must be struck with the wonderful extent to which human dexterity and ingenuity can be carried. In a more superstitious age, and while our statute books were filled with laws against magic and witchcraft, these gentlemen would have been in no very enviable situation, and though they distinctly disclaim all supernatural power, it is even now hard to disbelieve it; and we have no hesitation in placing Mr. Shaw in the very first rank of his profession.[123]

The performance of 'human dexterity and ingenuity' connected the display of automata to the objects of science and industry presented at the exhibition, so many of which would have appeared to possess 'supernatural power' as little as one generation previously.

Conclusion

The exhibition's extended closure mirrored its protracted installation. Just as it had opened before many of the exhibits were ready, so it closed by increments too. The committee sought this final extension to augment the profits intended to benefit the Dispensary and the House of Recovery. While it would have been embarrassing for the Mechanics' Institution to fail in its object to remove its debt through the profits from the first season, for the second season to be unsuccessful in its charitable aims added an additional moral dimension to the problem. The final figures appear not to have been published in the local periodical press, which for an event so well-documented throughout its run suggests that the profits were not high enough to be celebrated. The annual report of the Leeds Mechanics' Institution and Literary Society for 1843 offered the following evaluation of the exhibition, which is worth reproducing in full:

[123] 'Leeds Public Exhibition', p. 5. By 1849 Shaw was performing without D'Auvigney, giving lectures at the Royal Polytechnic Institution in London on 'The Art of Magic illustrated and explained'. See 'Royal Polytechnic Institution', *Mechanics' Magazine, Museum, Register, Journal and Gazette*, 6 January–30 June 1849, p. 216.

> In compliance with the resolution of the last General Meeting, your Committee took immediate steps for carrying this arduous undertaking into effect. They invited the co-operation of all those who had taken an active part in the last one; and though they had to lament the falling off, from various causes, of some former friends, they met with a large share of support; and they and the Society at large can never feel too deeply grateful to those Gentlemen, who, unconnected with this Institution, gave their most valuable aid in furtherance of it. Their names must ever be ranked amongst its benefactors. The official accounts not having been made up, your Committee are not enabled to lay before you the exact results. But as far as can be ascertained, it appears that there will not be realised to this Institution much more than £400. The receipts are £2,150, expenses £1,700. When the unparalleled exertions and indefatigable labours of a number of gentlemen during so long a period are looked at, this result will no doubt appear a small one, and be disappointing to many. Into its causes it is not necessary further to enter. The amount realised will nevertheless form a most valuable addition to the funds of the Society. And if the results in a pecuniary point of view have not been such as might have been wished for, in most other respects which recommend institutions of this nature, it has been full of happy and satisfactory results, and your Committee have no cause to regret having embarked on it.[124]

The profit amounted to half the £800 debt borrowed against the Institution's building, and as such the fundamental objective of the exercise—to clear this debt—had not been met.[125] The total receipts for the 1839 exhibition had been £3,407 17s. 2½d. with a profit of £1,780.[126] The receipts for 1843 were down by over £1,000 but the profits were over four times smaller, suggesting that the expenses for the 1843 exhibition

[124] Leeds Mechanics' Institution and Literary Society, *Report of the Committee of the Leeds Mechanics' Institution and Literary Society, presented to the annual meeting on the 18th January, 1844, with a list of the Officers and Committee for 1844, &c. &c.* (Leeds: printed by Edward Baines and Sons, 1844), p. 14.

[125] Edward Baines, *Report Presented to the First General Meeting of the Leeds Mechanics' Institution and Literary Society* (Leeds: printed by Edward Baines and Sons, 1842), p. 7.

[126] John Mayhall, *The Annals and History of Leeds, and other places in the County of York, from the earliest period to the present time* (Leeds: Joseph Johnson, 1860), pp. 461–62.

had been far greater. That the committee chose to pass over the causes of the reduction in both the receipts and the profits indicates a level of embarrassment with their diminished success, although they were keen to emphasise the moral and intellectual benefits of their endeavour.

The decision to stage an exhibition in 1843 was based in part on the assumption that the success of the Leeds Public Exhibition of 1839 could be matched and, ideally, surpassed. The projection was based on the involvement of many of the same people, a similar charitable cause—to benefit the Leeds Mechanics' Institution and Literary Society—and a sense of confidence that the town and its environs would provide an even stronger collection of objects now that the model was proven. The *Leeds Mercury* and Edward Baines Junior continued to champion the event, while the rival *Leeds Times* was more cautiously optimistic and attentive to the more challenging economic context. Civic rivalry continued to be a significant motivation, as the major towns to the east and west—Hull, Liverpool and Manchester—had recently staged comparable exhibitions and Leeds could no longer rest on its success four years previously to maintain its position.

While the 1843 exhibition contained broadly similar categories of objects—loaned by broadly similar upper- and middle-class constituencies—as its predecessor, there were significant differences between the two. The beginning and end of the 1843 exhibition were less rigidly defined, with loans and performances added and subtracted throughout to encourage return visits and the purchase of season tickets. The overall display remained polytechnic, although works of fine art were no longer presented so closely alongside industrial machinery, which was displayed in a separate annex. The 1843 exhibition demonstrated an increase in historical consciousness through its 'Museum of Antiquities', which was staged to evoke a generalised sense of the past through an arrangement of objects associated with aristocracy. The place of portraits of local and national worthies was more prominent, intended to provide object lessons in religious, political and scientific achievement. In many ways this exhibition established a pattern that would be repeated in Leeds over the course of the century, with organisers ever increasing their ambitions and expenses. Although the resulting balance sheet rarely provided the projected returns in bare financial terms, it almost always provided the impetus to try again.

3

The Third Leeds Polytechnic Exhibition, 1845

A little over a year after the final closure of the Second Leeds Public Exhibition of 1843, the *Leeds Mercury* reported the intention to hold a Third Leeds Polytechnic Exhibition in the summer of 1845. The Leeds Mechanics' Institution and Literary Society began the year £600 in debt—£200 less than it had been ahead of the previous exhibition.[1] Unlike the exhibitions of 1839 and 1843, however, the profits generated by the third iteration were intended to contribute to the development of public walks and baths rather than to relieve the Mechanics' Institution of its liabilities. There were to be four trustees tasked with the organisation: the Mayor of Leeds, the Vicar of Leeds and the two Members of Parliament for the borough.[2] A meeting of these 'highly respectable gentlemen' took place on 27 January 1845 at the Philosophical Hall. The Vicar of Leeds, Walter Farquhar Hook (1798–1875), presided over the meeting and referred to a recent parliamentary grant in support of public walks and baths, alongside £30,000 raised in subscriptions for the same purpose in Manchester, with Birmingham and Sunderland also in pursuit of this end.[3] Discontinued for the Second Public Exhibition in 1843, the term 'polytechnic' re-emerged as a descriptor of the Third

[1] Leeds Mechanics' Institution and Literary Society, *Report of the Committee of the Leeds Mechanics' Institution and Literary Society, presented to the annual meeting, on the 22nd January, 1845, with a list of the officers and committee for 1845, and an addition Supplement to the Catalogue, containing the additions to the library from Jan. 1843, to Jan. 1845* (Leeds: printed by Edward Baines and Sons, 1845), p. 4.

[2] 'Public Walks and Baths', *Leeds Mercury*, 25 January 1845, p. 4; 'Public Walks and Baths for the Working Classes', *Leeds Times*, 25 January 1845, p. 5.

[3] 'Proposed Polytechnic Exhibition in Leeds', *Leeds Mercury*, 1 February 1845, p. 1; 'Public Baths and Walks:—Proposed Exhibition in Leeds', *Leeds Times*, 1 February 1845, p. 5; 'Public Baths and Walks in Leeds.—Polytechnic Exhibition', *Leeds Intelligencer*, 1 February 1845, p. 5.

Exhibition from these earliest meetings, which suggested that the organisers thought it might attract higher numbers of visitors. Emerging technologies were certainly at the forefront of their minds when the committee set up a competition for the best models of steam engines and original mechanisms, urging that 'practical mechanists who would like to try their skill, should lose no time in preparing'.[4]

The committee made their first public appeals for donations of objects for display in the *Leeds Mercury* and the *Leeds Times* on 15 February 1845 and in the *Leeds Intelligencer* on 22 February 1845.[5] The committee articulated the achievements of the previous two exhibitions and used them to reassure potential donors that their possessions would come to no harm:

> The growing taste of the Public for intellectual recreation has been abundantly proved by the gratifying results of two former Polytechnic Exhibitions in this Town. These Exhibitions were visited by many thousands of persons, and the preservation from injury of the assemblage of valuable Specimens comprised in them, amply demonstrates that their value and worth were fully appreciated.[6]

The advertisement made it clear that the primary objective was the improvement of sanitary conditions for the working classes, with attendant benefits 'in improving health, in giving an opportunity for agreeable recreation, and in withdrawing the Labourer from the temptations of the Dram-shop to amusements which can be shared by his Family, and will not be followed by poverty, disease, or crime'.[7] These beliefs would come to be more famously articulated by Henry Cole when in November 1857 he expressed the opinion that 'the evening opening of Public Museums may furnish a powerful antidote to the gin palace'.[8]

The urban centre of Leeds had become a contested space by the mid-nineteenth century; what had been constructed as an affluent development around Park Square on the edge of the town during the eighteenth century had been subsumed by the urban industrial centre by the nineteenth. The factories around Wellington Street to the west,

4 'Proposed Polytechnic Exhibition', *Leeds Intelligencer*, 8 February 1845, p. 5.

5 See, for instance, 'Leeds Polytechnic Exhibition, for providing Public Walks and Baths', *Leeds Mercury*, 15 February 1845, p. 4.

6 'Leeds Polytechnic Exhibition, for providing Public Walks and Baths', p. 4.

7 'Leeds Polytechnic Exhibition, for providing Public Walks and Baths', p. 4.

8 Henry Cole, *The Functions of the Science and Art Department* (London: Chapman and Hall, 1857), p. 26.

the densely populated ward of Kirkgate to the east and the heavy industry south of the River Aire had been associated with outbreaks of cholera and other contagious diseases in an influential report by Robert Baker published in 1842.[9] Baker mapped these outbreaks against the areas populated by the working classes and the 'less cleaned districts', and noted a strong correlation between poverty, industry and disease.[10] Baker summarised the social geography of Leeds in the starkest terms:

> by drawing a line through the centre of the map from north to south, the deaths in proportion to population on the east side of the map were, in 1839, as 1 to every 24; while on the other hand, in those parts of the town where the streets are spacious and wide, and the drainage sufficient, the deaths were only 1 to 36; both ratios being exceedingly high, but the difference remarkable.[11]

The exhibition was tied into these contemporary discussions about public health and sanitary improvements and aligned with a visit from James Simpson organised by the Mechanics' Institution.[12] Simpson's lectures took place at Christ Church School Room on Meadow Lane on 8 and 9 April 1845. The Mechanics' Institution emphasised that their primary audience was intended to be the working classes and that the lectures would be particularly accessible and relevant to their lives.[13] That the events took place not in the lecture theatre of the Mechanics' Institution or the Philosophical Hall, but in a space usually used as a Sunday School among the factories of industrialised Hunslet, south of the River Aire, demonstrated a commitment to this aim. Reserved seats were available from the offices of the *Leeds Mercury*, *Leeds Times* and *Leeds Intelligencer* at 6d. each and admission was free to the body of the room.[14] The mayor Darnton Lupton (1806–73) presided over the lectures in his capacity as a member of the committee of the Third Polytechnic Exhibition and one of its four trustees. It is not clear whether the purpose of the lectures was to educate the working classes about measures they could take to improve the sanitary conditions of their own lives, or to inform them for the purpose of advocating for

9 Robert Baker, *On the State and Condition of the Town of Leeds in the West Riding of the County of York* (London: Her Majesty's Stationery Office, 1842).

10 Robert Baker defined the working classes as those living in houses where the annual rent was below £10.

11 Baker, *On the State and Condition of the Town of Leeds*, p. 19.

12 'Public Baths and Walks', *Leeds Intelligencer*, 29 March 1845, p. 5.

13 'Polytechnic Exhibition', *Leeds Times*, 5 April 1845, p. 5.

14 'Polytechnic Exhibition', p. 5.

improvements to be made from above, or simply to articulate the nature of the problem. The committee may have been keen to align themselves with the event as a means of priming their potential audience to support the exhibition and its objectives.

Same difference

Meanwhile the committee's belief that 'a vast store of Specimens and Works of Art of the highest order are yet available' began to take tangible form.[15] In April 1845 the committee announced some of the works of art already secured:

> The Committee of the forthcoming Exhibition have promised to them the use of eleven Cartoons, three of which were successful in the recent competition for the Parliamentary premiums. There are not of the same collection as those which have already been exhibited in Leeds, but are alike universally admitted to be pre-eminent specimens of art. The Committee have also succeeded in obtaining the loan of numerous groups of sculpture, amongst which are several of those exhibited not very long ago in Westminster Hall, one of which, executed by Thomas Earle, Esq., a native of Hull, obtained the gold medal prize at the meeting of the Royal Academy. A great variety of pictures have also been secured for the exhibition, amongst which is Haydon's *chef d'œuvre*, "Marcus Curtius," and also some of the best productions of Landseer, Sidney Cooper, Martin, Rembrandt, &c.[16]

It was common for these pronouncements to qualify the objects with reference to the previous two exhibitions. The committee demonstrated that the works of art to be displayed at the 1845 exhibition were related to what had already been seen in Leeds—the cartoons for instance—but crucially that the specific examples would be new to the town. This provided audiences with an incentive to visit: it reassured them that what they enjoyed about their experiences would be replicated, but not so closely as to merely repeat the exercise. The balance between

[15] 'Leeds Polytechnic Exhibition, for providing Public Walks and Baths', Correspondence, 1–19 April 1845. British Library, MS Papers of the Prime Ministers of Great Britain, Series Two: Sir Robert Peel: Part Five: Additional Manuscripts 40557-40581 40564, p. 223.

[16] 'Prospect of the Leeds Exhibition', *Leeds Intelligencer*, 12 April 1845, p. 5.

sameness and difference is an important consideration as it points towards one of the reasons that the 1845 exhibition would be the last to follow this model.

The committee reported that they were ready to receive donations for display in April 1845 and stated their ambition to open to the public during the following month.[17] The committee used the local periodical press not only to generate interest and excitement about the content, but also in the people who had contributed loans. For instance the Prime Minister, Sir Robert Peel, was reported to have 'signified his intention of favouring the exhibition by the loan of some paintings from his splendid collection'.[18] The committee were keen to demonstrate that support was not restricted to local people and could be found at the highest levels of British society: 'The Earl of Zetland, though not personally connected with Leeds, has offered to lend a specimen of Guido, of immense value.' Thomas Dundas, 2nd Earl of Zetland (1795–1873), may not have been directly connected to Leeds, but his family home was Aske Hall near Richmond, he served as a Member of Parliament for Richmond and later York, and had been appointed Lord Lieutenant of the North Riding of Yorkshire in 1838, so his connections to Yorkshire were substantial. Most symbolically significant of all was the news in May 1845 that Prince Albert 'has been pleased to evince his approbation of the object which the committee have in view, by consenting to accept the office of patron of the exhibition'.[19]

Towards the end of May 1845 the *Leeds Intelligencer* reported that the committee had made good progress and that paintings by Rembrandt, Gainsborough and Martin had already been unpacked. The newspaper foregrounded the quality of sculpture that would be displayed, alongside the construction of an atmospheric railway, 'a number of highly interesting optical contrivances', dissolving views and chrometropes. The *Intelligencer* expressed confidence that the resulting exhibition would have 'every chance of carrying out the noble object to which its proceeds will be applied,—those of providing, especially for the working classes, but likewise for all, the sources of pure and elegant recreation in public walks, and of salubrious cleanliness in cheap baths'.[20] As had been the case during the preparations for the Second Public

17 'Leeds Public Walks and Baths.—Polytechnic Exhibition', *Leeds Times*, 26 April 1845, p. 5.

18 'Leeds Public Walks and Baths.—Polytechnic Exhibition', p. 5.

19 'Leeds Polytechnic Exhibition for Providing Walks and Baths', *Leeds Times*, 10 May 1845, p. 5.

20 'Leeds Polytechnic Exhibition', *Leeds Intelligencer*, 17 May 1845, p. 5.

Exhibition in 1843, the committee had overextended itself by aiming to open to the public by the end of May 1845. On 24 May an opening date of 2 June was advertised in the *Leeds Times*.[21] One week later the *Leeds Mercury* reported that the exhibition would open on Monday 9 June.[22] While such a delay would represent a considerable embarrassment and have significant implications in present-day exhibition making, the committee and their objectives were so strongly supported by—and embedded in the social, political and economic networks of—the local periodical press, that it was never presented as a failing or a cause for concern. In much the same way as the comparable postponed opening had been reported in 1843, the delay was framed as a reflection of the support generated for the project and the commitment of the committee to its realisation. 'So many additions have been made to the stores', the *Leeds Times* reported, that 'in order to perfect the arrangements' the only solution was to defer the opening date.[23]

Diminishing returns, increasing costs

The committee expressed other concerns and intimated that local support in the form of loans had not been as strong as they had hoped. They reminded prospective lenders that Leeds was the 'fourth town in wealth and importance in the British empire' and as such, its local population were expected to contribute to the exhibition with objects of commensurate status. They reassured—and implicitly threatened—their benefactors with the postulation that if Leeds increased its provision, the town would 'fear no superiority on the part of Liverpool or Manchester, or even London'.[24] That the committee called for further support at this late stage suggests a certain amount of anxiety about their prospects. It was clear that for those with a stake in this ongoing cultural competition, they did fear the superiority of their regional neighbours. That the organisers also sought to compare their efforts with exhibitions in London demonstrated a real belief that the combined resources of the regional gentry and manufacturing elite could bring together a collection to rival any in the country. The

21 Advertisement and Notices, *Leeds Times*, 24 May 1845, p. 1.
22 'Leeds Polytechnic Exhibition, for providing Public Walks and Baths', *Leeds Mercury*, 31 May 1845, p. 4.
23 'Polytechnic Exhibition', *Leeds Times*, 31 May 1845, p. 4.
24 'Leeds Polytechnic Exhibition, for providing Public Walks and Baths', *Leeds Times*, 24 May 1845, p. 5.

London-based *Art-Union* journal also made reference to the prosperity of the town in its notice on the forthcoming exhibition:

> We direct attention to an advertisement which announces an approaching Exhibition in this wealthy town—the commercial capital of Yorkshire. The object in view is an excellent one, and the artists who contribute works will, while increasing their own fame and advancing their own interests, have the satisfaction to know they are aiding a good cause, and extending a knowledge and love of art.[25]

The cost of a day ticket remained the same as in 1839 and 1843 at 6d. each. The cost of a season ticket, however, rose to 5s.[26] An unnamed member of the Leeds Mutual Improvement Society wrote to the *Leeds Times* to complain about the increase:

> I have seen with surprise, that the Committee have fixed the price of Season Tickets at 5s. This is a matter of very deep regret with working men, who cannot afford to spend 5s. out of their scanty earnings; whereas, if the price was the same as at the former exhibitions in this town,—namely, 2s. 6d., hundreds of working men would avail themselves of it to spend their evenings in an instructive and rational manner. As it is, they will not be able to afford it; especially if the wages of the men be only very trifling. I feel persuaded, that if the Committee would go upon the principle of small profits, and quick returns, they would realise a considerably greater amount of money, and also do proportionally greater good.[27]

Although the *Leeds Times* endorsed this view, the committee initially stood firm. In the first week it was reported that season tickets were selling well despite 'An impression that the price would be lowered', which, 'has rather had a tendency to prevent people from buying, but we cannot hear that any such reduction is contemplated'.[28] The *Leeds Intelligencer* carried the same message, with the further detail that 800 season tickets had been sold up to 14 June 1845.[29]

[25] 'Leeds', *Art-Union*, 1 April 1845, p. 102.

[26] 'Leeds Polytechnic Exhibition, for providing Public Walks and Baths', *Leeds Mercury*, 31 May 1845, p. 4.

[27] To Readers and Correspondents, *Leeds Times*, 31 May 1845, p. 4.

[28] 'The Leeds Polytechnic Exhibition', *Leeds Times*, 14 June 1845, p. 5.

[29] 'Leeds Polytechnic Exhibition', *Leeds Intelligencer*, 14 June 1845, p. 5.

However, reports that the price would stand did not settle the matter. One month later the *Leeds Times* published a second complaint from a correspondent referred to as 'C.', who wrote:

> the Committee have shut the very men, for whose benefit it professed to be, by fixing the price of the season tickets at 5s. each. Very few working men indeed can conveniently give 5s., while nearly all might give 2s. 6d.: and this is to make the company 'select,' and to avoid the uproars that were created last Exhibition. It is well-known, however, that those disturbances were not caused by the working people, but by those who called themselves 'gentlemen,' privileged, they thought, to have some fun at other people's expense. And now there is a regulation made purposefully for the rich: "a family taking two season tickets at 5s. each, may have tickets at 2s. 6d. for their children, and one of an *attendant* [original emphasis], at the same price." Every one must see the injustice of this regulation. A favour is extended to the families of the wealthy classes, which is denied to the working men. Is this fair—is it in consistence with the professed objects of the Exhibition?[30]

The letter contained several interconnected grievances. First, the same complaint we have already heard—that the rise in cost of a season ticket to 5s. was not affordable for the working classes and therefore excluded the people whom the exhibition was designed to benefit. The author then accused the committee of raising the price in order to deliberately limit access, citing unspecified 'uproars' at the exhibition of 1843. It is not clear to what exactly the author referred, as the local press—generally supportive of the exhibitions and those associated with them—only reported a couple of minor instances of petty theft as detailed in the previous chapter. Lastly, the complainant drew attention to the perceived injustice of allowing those families who could afford to purchase two season tickets at the full price to purchase further tickets for children and an attendant at the original price of 2s. 6d.

The *Journal of the Leeds Polytechnic Exhibition* reported an incidence of comparable bad behaviour that took place on 2 July 1845, which suggested that the perceived intention to regulate the conduct of visitors through economic exclusion had not been entirely successful:

30 'Polytechnic Exhibition', *Leeds Times*, 12 July 1845, p. 4.

> The Curators regret to say, that on Wednesday night they had to exercise a very unpleasant duty to the Exhibition. A number of well dressed "young gentlemen," (?) were behaving in a most disorderly and riotous manner; two of the most prominent were walked into the Committee room, and as the facts were strong against them, their tickets were forfeited. The Curators cannot allow the Visitors to be disturbed by any such nuisance as was frequently complained of at the last Exhibition. In future, parties will be in danger of having their names exposed, as well as their ticket being forfeited.[31]

The report supported the previous claim that it was not primarily the working classes whose behaviour at the exhibition was disruptive. The opposition to the increased cost coupled with disappointing visitor figures during the first month provided the committee with enough cause to reconsider, as the following week brought the news that season tickets had been reduced to the same price as the 1839 and 1843 exhibitions at 2s. 6d. for the remainder of the season, and working-class school pupils would be able to visit for 1d. each on Fridays.[32] The price of tickets, the narratives constructed around the exhibition and the penalties for transgression were mechanisms to modify the behaviour of visitors through what Morris has described as 'the organisation of consent'.[33] The committee performed their authority, in Morris's terms, 'through ritual, through politics and ideology' to influence their audience, rather than the blunt imposition of social control.[34]

The Hall of Sculpture

Despite the delay in opening to the public the local periodical press had again been invited to preview and report on the exhibition before it was complete. The *Leeds Mercury*, *Leeds Times* and *Leeds Intelligencer* all published extensive reviews in their editions of 7 June 1845 ahead of the public opening on 9 June 1845. The *Mercury* introduced its report with the qualification that: 'though the arrangements were necessarily somewhat incomplete, we are nevertheless enabled to form an opinion

31 *Journal of the Leeds Polytechnic Exhibition, for providing Public Walks and Baths*, 5 (7 July 1845), p. 38.

32 'Leeds Polytechnic Exhibition', *Leeds Intelligencer*, 19 July 1845, p. 5.

33 R.J. Morris, 'Governance: Two Centuries of Urban Growth', in *Urban Governance: Britain and Beyond since 1750*, ed. by R.J. Morris and Richard H. Trainor (Aldershot: Ashgate, 2000), pp. 1–14 (p. 8).

34 Morris, 'Governance: Two Centuries of Urban Growth', p. 10.

of its merits. It contains a vast and splendid collection of all that is truly beautiful or novel, curious, useful or valuable in science and art.' In comparison with the two previous exhibitions, it was reported that 'it is equally large in its extent,—perhaps not less diversified in its attractions,—and is much superior to either in a scientific point of view.[35] In contrast to the previous iterations, however, the objects that visitors first encountered on entry were works of art. A further point of distinction was that these works of art were not paintings, prints or drawings, but sculptures. For the first time an exhibition included a substantial and significant collection of contemporary and historical sculpture as a discrete display in what had been the Museum of Antiquities in the Ante-Room. The *Leeds Mercury* described it as 'a very auspicious and attractive commencement'.[36] The Hall of Sculpture received a positive assessment in the *Leeds Times* too:

> This forms, what may be called, the grand feature of distinction betwixt the present and preceding exhibitions. On former occasions, the collection of sculpture was miserably meagre. Now we have a room almost exclusively devoted to it. The floor is covered with groups and figures of statuary; while busts of distinguished personages crowd the tables on every side... This room is decidedly one of the most interesting and novel in the Exhibition.[37]

The *Leeds Intelligencer* concurred and provided a more florid introduction to the new attraction:

> And here are displayed some of the most exquisite productions of the most eminent artists. The statesman and the poet, the philosopher and the divine, hold their stations here by proxyship of marble, and the feeling of veneration and respect for the departed great ones who seem to be gazing upon you with the intensity of animation, will intrude upon your soul and make you feel subdued and reverent while in so illustrious a companionship. The statuary includes several excellent classical illustrations of the highest order of modelling.[38]

35 'Leeds Polytechnic Exhibition', *Leeds Mercury*, 7 June 1845, p. 5.
36 'Leeds Polytechnic Exhibition', p. 5.
37 'Leeds Polytechnic Exhibition', *Leeds Times*, 7 June 1845, p. 5.
38 'Leeds Polytechnic Exhibition', *Leeds Intelligencer*, 7 June 1845, p. 5.

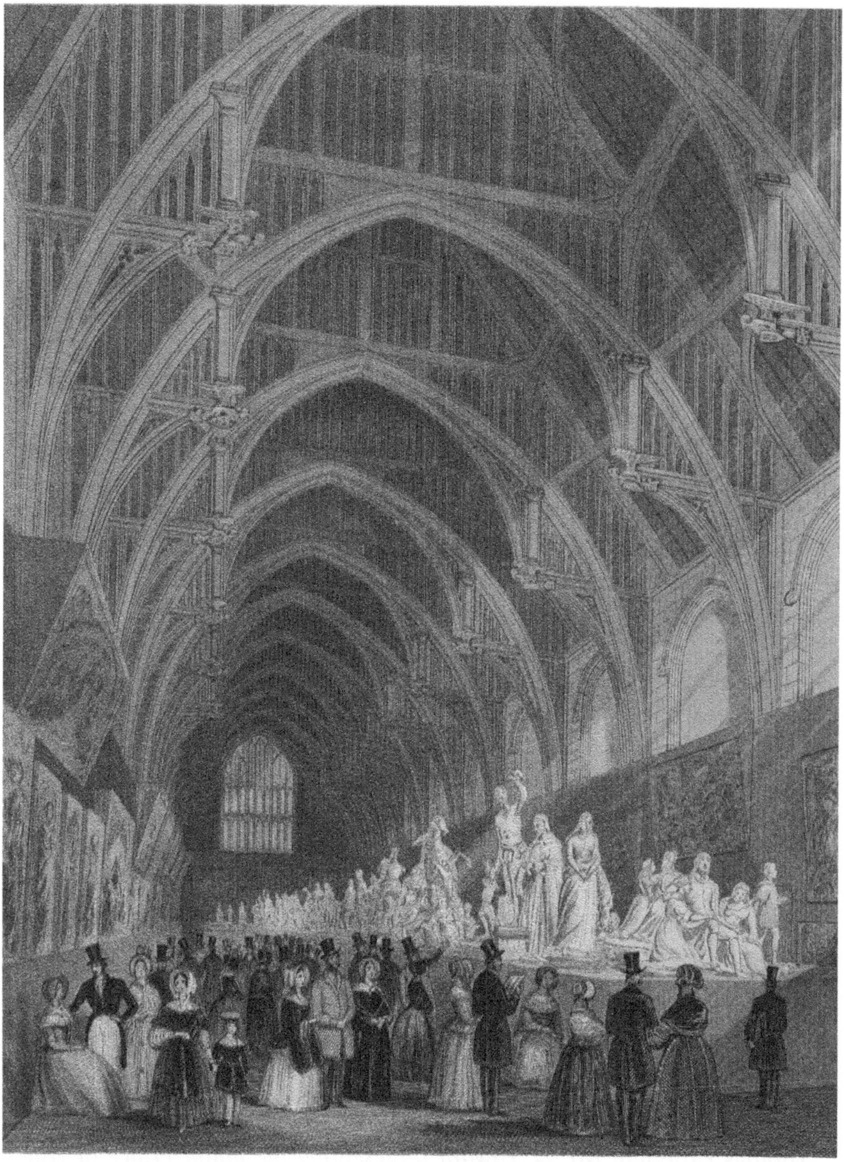

Figure 3.1 Edward Radclyffe after Benjamin Sly, *Public Exhibition of Frescoes & Sculpture in Westminster Hall*, 1844, steel engraving. Collection of the author

The sources of these objects were both national and local in origin (Figure 3.1). Five of the sculptures on display in Leeds were shown as part of an exhibition at Westminster Hall in London in 1844.[39] The exhibition was held in association with the Royal Fine Arts Commission whose task it was to commission painting and sculpture as part of the rebuilding of the Houses of Parliament. The Commissioners invited sculptors to submit up to two models as examples of their work, which could either have been made especially for the occasion or during the previous five years. The competition was open to British artists and to artists who had lived in Britain for more than ten years. It was not intended as a means to select from the models works to be realised in bronze or marble, but rather a way of choosing a pool of sculptors who might be tasked with producing decorative statuary for the new building.[40] Two models by the Hull-born sculptor Thomas Earle (1810–76) were displayed at Westminster Hall in 1844 and then in Leeds in 1845.[41]

The art critic Henry G. Clarke wrote and published both a handbook to the exhibition and, shortly afterwards, a longer 'critical examination'. Of the sculpture in general he celebrated the opportunity to consider so many works of contemporary British sculpture together:

> The double line of statues down the middle of the magnificent Hall, consists of a large portion of the best works that have been crowded and ill-seen in the sculpture-room of the Royal Academy during the last five years, and others of a more national character executed for this especial occasion: there never was so good an opportunity of judging British talent in this department of art.[42]

[39] The five sculptures were Thomas Earle's *Edward I. presenting the Infant Prince of Wales to his Welsh subjects* and *An Ancient Briton defending his Wife and Children* (both contributed by the artist and not for sale); William Scoular's *Adam and Eve lamenting over the dead body of Abel* and *Shepherd Boy* (both contributed by the artist and listed for sale); and William Spence's *Theseus killing the Minotaur* (contributed by the artist and listed for sale).

[40] See Benedict Read, *Victorian Sculpture* (New Haven, CT: Yale University Press, 1982), pp. 82–84.

[41] Thomas Earle lent No. 23, *Edward I. presenting the Infant Prince of Wales to his Welsh subjects*, and No. 28, *An Ancient Briton defending his Wife and Children*, neither of which were listed for sale. See Edward Hailstone, ed., *Catalogue of the Leeds Third Polytechnic Exhibition, for Providing Public Walks and Baths, containing a magnificent collection of Sculpture, Cartoons, Paintings, Antiquities, Curiosities, Philosophical Apparatus, Mechanical Models, &c. with the names of the contributors, opened June 9th, 1845, at the Music-Hall* (Leeds: printed by Bolland and Kemplay, 1845), p. 5.

[42] Henry G. Clarke, *A Hand-Book Guide to the Cartoons, Frescos, and Sculpture,*

In the handbook he wrote of Earle's *An Ancient Briton defending his Wife and Children*: 'A good intention well executed but rather imperfectly composed, the huddling of the mother and children is neither well connected with the father, nor contrasted by a bold upright attitude in him: it is undermined'; and of *Edward I. presenting the Infant Prince of Wales to his Welsh subjects* he surmised: 'Not like the talented, unrelenting Edward, but the child and his manner of holding it are pretty'.[43]

In his *Critical Examination* Clarke entered into further discussion of the perceived merits and deficiencies of *An Ancient Briton protecting his Family*:

> The intention and execution of this group are generally interesting, yet the forms are too plump, the features and other details deficient in delicate variety; they are all alike. The action of the father is undecided, as he neither groups with the other figures nor preserves the upright bold bearing of their protector; his face is also wanting in determination.[44]

Clarke was also critical of Earle's second model, *Edward I. presenting the first Prince of Wales*, which he described as: 'A figure in armour holding a baby in his arms, which he does rather gracefully; but it does not seem to us like the determined, unrelenting Edward holding up his infant child to the abused Welchmen [*sic*]. It is rather clever, not characteristic.'[45] Both models depicted fathers and their children and both were criticised for lacking in 'determination'; perhaps used in this context as a synonym for a certain conception of masculinity and the conventions of its expression in sculpture.

Earle's two models received a more positive reception in Leeds. The *Leeds Mercury* described their placement as 'the most prominent objects of sculpture... "The Briton protecting his wife and family," by Earle, occupying the centre of the room; and "Edward the 1st presenting the young Prince of Wales to the Welsh people," also by Earle, which

Designed in pursuance of Notices issued by Her Majesty's Commissioners on the Fine Arts, in furtherance of their object for ornamenting the New Palace at Westminster. And now exhibiting in Westminster Hall (London: H.G. Clarke & Co., 1844), p. iii.

43 Clarke, *A Hand-Book Guide to the Cartoons, Frescos, and Sculpture*, p. 27.

44 Henry G. Clarke et al., *A Critical Examination of the Cartoons, Frescos, and Sculpture, exhibited in Westminster Hall* (London: H.G. Clarke & Co., 1844), p. 48.

45 Clarke et al., *A Critical Examination of the Cartoons, Frescos, and Sculpture*, p. 48.

stands boldly out at the extremity of the hall.'[46] The *Leeds Times* called the former 'a fine model group'.[47] The *Leeds Intelligencer* reported too on the prominence and qualities of Earle's work: 'One of the principal groups of model-sculpture represents an Ancient Briton defending his family, by Earle. "Edward the First presenting his infant son to the Welsh people," forms another attractive specimen, by the same artist. The figure and demeanour of Edward are full of natural nobility, and contrast well with the smiling innocence of the naked child.'[48] That the works were both described as models and as particularly prominent in the arrangement suggests that they were full-size works in plaster, awaiting a state commission that would never come. The *Intelligencer* published a follow-up piece the next week and again offered a favourable commentary on Earle's sculptures:

> We next observe that noble figure of a stalwart fellow standing in a position of valorous defence, and beneath his well-placed shield we see the figures of a woman and two children whose looks evince a world of anxiety and alarm. That is "An Ancient Briton defending his wife and children," and none who see it can fail to admire the spirit which pervades this group, as well as the anatomical accuracy which distinguishes the principal figures. It is the production of Mr. T. Earle, a name which in itself is "a tower of strength." Other statues by this eminent artist are also in the collection, such as "Edward I. presenting the infant Prince of Wales to his Welsh subjects," and "Hercules protecting Hesione from the Sea Monster".[49]

The sculptures from the Westminster Hall exhibition were joined by six cartoons that had also been displayed in London for the same purpose—the ornamentation of the new Houses of Parliament. The *Leeds Intelligencer* celebrated these examples as 'magnificent specimens of the class of art to which they belong, and manifest great powers of historical design by some of our living artists'.[50] Three of the six cartoons displayed in the Hall of Sculpture were awarded fourth class prizes at the Westminster Hall exhibition in 1843: *St. Augustine proclaiming Christianity to the Britons*, by W.C. Thomas; *The Seven Acts of Mercy*,

46 'Leeds Polytechnic Exhibition', *Leeds Mercury*, 7 June 1845, p. 5.
47 'Leeds Polytechnic Exhibition', p. 5.
48 'Leeds Polytechnic Exhibition', *Leeds Intelligencer*, 7 June 1845, p. 5.
49 'Leeds Polytechnic Exhibition', *Leeds Intelligencer*, 14 June 1845, p. 5.
50 'Leeds Polytechnic Exhibition', *Leeds Intelligencer*, 7 June 1845, p. 5.

by E.V. Rippingale; and *The Brothers driving out Comus*, by John Green Waller.[51] They were hung high on the walls of the Ante-Room, beneath which bas reliefs, medallions and other ornamental carvings were suspended. Statuettes were shown on tables and included works in ceramic, bronze, ivory and marble with literary, poetic and religious themes.[52] In addition to objects carved and cast in these traditional materials, a cast iron statuette 'of chaste workmanship' from the Hazle Foundry in Cornwall was also included. It was presented as an object that united art with industry, both in the method and context of its production and its purpose: 'that it was presented by the Royal Cornwall Polytechnic Society to Mr. Richard Hunt, of Devonport, on the exhibition of he newly invented calculating instrument'.[53]

Models of Samuel Nixon's (1803–54) personifications of the four seasons as children, presumably also presented in plaster at full-size, were displayed in the four corners of the Hall of Sculpture.[54] The original commission had come in the summer of 1839 through the architect Philip Hardwick, who employed Nixon to carve sculptural decorations for the new Goldsmiths' Hall in the City of London.[55] The final marble iteration of *The Four Seasons* was installed on the lower staircase by early 1844.[56] The prominent position of these sculptures and their part in a major new building ensured that they were noticed by contemporary commentators. The *Illustrated London News* dedicated a full page to the scheme, complete with wood engravings of each figure and an illustration of the staircase in context (Figure 3.2). It described the figures as 'a most artistical accession to its embellishment ... the poetic feeling displayed in the designs is ably carried out in the extreme

51 Henry G. Clarke, *A Hand-Book Guide to the Cartoons now exhibiting in Westminster Hall, Designed in pursuance of Notices issued by Her Majesty's Commissioners on the Fine Arts, in furtherance of the object of their enquiry, whether Fresco Painting can be applied with advantage to the decoration of the Houses of Parliament* (London: H.G. Clarke & Co., 1843), p. 9. See also Numbers 1–6 in Hailstone, ed., *Catalogue of the Leeds Third Polytechnic Exhibition*, pp. 3–4.

52 'Leeds Polytechnic Exhibition', p. 5.

53 'Leeds Polytechnic Exhibition', p. 5. Number 51 in Hailstone, ed., *Catalogue of the Leeds Third Polytechnic Exhibition*, p. 6.

54 Numbers 18–21 in Hailstone, ed., *Catalogue of the Leeds Third Polytechnic Exhibition*, p. 5.

55 'The Goldsmiths' Company', *Art-Union*, 15 June 1839, p. 88.

56 L.H. Cust and Terry Cavanagh, 'Nixon, Samuel (1803–1854), sculptor', *Oxford Dictionary of National Biography* (Oxford University Press, 2004), https://www.oxforddnb.com/view/10.1093/ref:odnb/9780198614128.001.0001/odnb-9780198614128-e-20213 (accessed 14 July 2020).

delicacy of their execution. They have been universally admired by the *cognoscenti* of the sculptural art; and will add materially to Mr. Nixon's well-earned fame.'[57] The *Leeds Intelligencer* confirmed that they 'have gained for him [Nixon] an exalted celebrity' and praised the work in the following terms: 'The symbolic character of each is finely pourtrayed [*sic*], and the transition of the seasons is developed with a poetic genius which does full justice to the delightful simplicity of nature'.[58] Alongside the display of sculptures and cartoons from Westminster Hall, Nixon's models of *The Four Seasons* connected Leeds to the latest activity in the visual arts in the capital.

The collection of portrait busts on display in the Hall of Sculpture attracted praise from the *Leeds Intelligencer*, which described them as 'all of the very first class'.[59] The subjects included the Bishop of Ripon by an unnamed sculptor; the Archbishop of York by Matthew Noble (1817–76); John Beckett by William Behnes (1795–1864); William Pitt by Joseph Nollekens (1737–1823); and Queen Victoria and Prince Albert by John Francis (1780–1861). It is worth considering one of the busts in particular because it represented one of the exhibition's four trustees and a member of the organising committee: Walter Farquhar Hook by William Day Keyworth (1817–97). This is also the only sculpture to have remained on display in Leeds. The collections of Leeds Museums and Galleries hold two marble busts of Walter Farquhar Hook by William Day Keyworth.[60] The only substantial difference between them is that one is signed and dated: 'W.D. Keyworth. Sculptor. Hull. 1844.' with an inset plaque on the pedestal that reads: 'This bust of Walter Farquhar Hook D.D. [Doctor of Divinity] Vicar of Leeds, was presented to Mrs. Hook in 1844, and with her permission transferred to the Victoria Hall on her husband's preferment to the Deanery of Chichester in 1859'. The sculptor was born in Hull in 1817 to a stonemason father who relocated the family to London to work under the eminent portrait sculptor Francis Chantrey (1781–1841). William worked under Chantrey's studio assistant Henry Weekes (1807–77) from 1831 to around 1834, when the family moved back to Hull. He retained links to London by exhibiting portrait busts at the Royal Academy from 1837 onwards. Looking at other examples of Keyworth's portrait busts—the majority of which were

57 'Goldsmiths' Hall.—Nixon's Statues of the Seasons', *Illustrated London News*, 13 January 1844, p. 21.

58 'Leeds Polytechnic Exhibition', *Leeds Intelligencer*, 14 June 1845, p. 5.

59 'Leeds Polytechnic Exhibition', p. 5.

60 Accession numbers LEEAG.SC.2005.0506 (the unsigned version in storage at Leeds Museums Discovery Centre) and LEEAG.2005.0506 (the signed version on display in the entrance hall of Leeds Civic Hall).

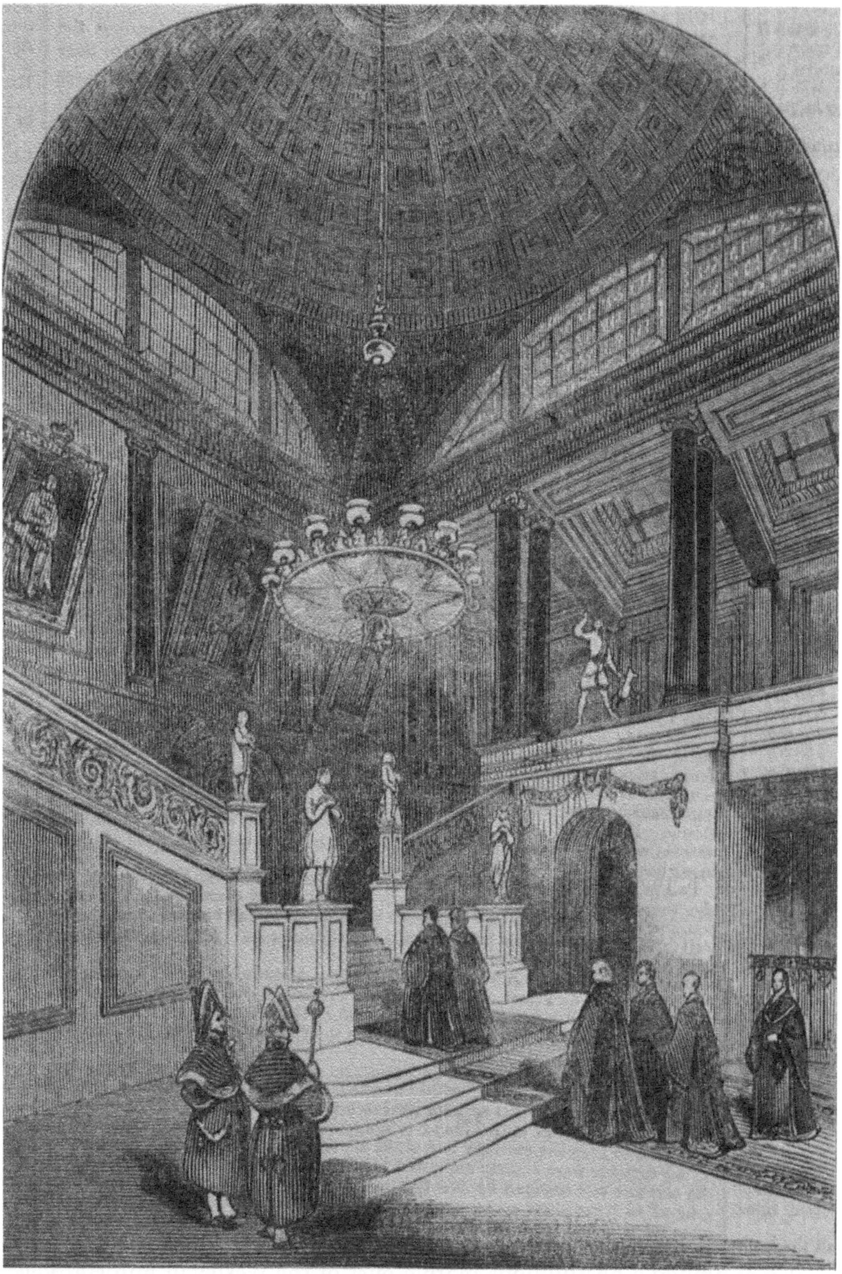

Figure 3.2 Anonymous, *Grand Staircase and Vestibule of Goldsmiths' Hall*, from *The Illustrated London News*, 13 January 1844, wood engraving. Collection of the author

made for civic buildings in his native Hull—his sculptural solutions to the depiction of the mid-nineteenth-century political, industrial and intellectual elite fall into two distinct types: those which represent the sitter in contemporary dress and those which adopt classicised drapery. It is very difficult to know whether the choice of garment belonged to the sitter, sculptor or patron. The vaguely Roman drapery used for the bust of Hook was perhaps a more unusual choice for a High Church Anglican, despite being an entirely conventional formal device for the period and context. It does not depict him in a cassock nor include any other external signifiers of his vocation. By stripping away these specific references, the object and its subject could transcend time and place and more readily be absorbed into a pantheon of great men.

The first reference to a bust of Hook by Keyworth comes in 1842, when the *Leeds Mercury* reported: 'We have seen a bust of the Vicar of Leeds, modelled by a young sculptor from Hull, Mr. W. D. Keyworth, which is an admirable likeness, and does great credit to the artist. The head is full of character and expression.'[61] The short notice does not record that the bust had been exhibited before, nor that it had yet been carved in marble. Given that the marble was not exhibited at the Royal Academy until 1844, it seems likely that the bust to which the *Leeds Mercury* referred was a preparatory work in clay or plaster, possibly shown at his Baker Street workshop in Hull to solicit support for the production of a marble version.[62] By exhibiting the marble bust in 1844 Keyworth consolidated the status of Hook at the most significant point in his career. Writing after Hook's death, his son-in-law described his position in 1844: 'the influence which the Vicar had acquired over all classes of people in Leeds, and the work which he had achieved, exceeded the most sanguine expectations of himself and his friends ... he seemed almost at the zenith of his power. He was at peace in his parish.'[63] Although this biography is hagiographic, the admiration Hook enjoyed in 1844 was confirmed when in November of that year the bust was presented to Hook's wife Anna, paid for by 200 subscribers 'as a testimony of esteem'.[64] This act of generosity was in sharp contrast to the open hostility faced by Hook immediately before his election

61 'Bust of the Rev. Dr. Hook', *Leeds Mercury*, 8 October 1842, p. 5.

62 Royal Academy of Arts, *Exhibition of the Royal Academy. MMDCCCXLIV. The Seventy-Sixth* (London: W. Clowes and Sons, 1844), pp. 51–52. Keyworth also exhibited a posthumous marble bust of the Wesleyan minister Revd J. Galland. Catalogue numbers 1313 and 1333 respectively.

63 W.R.W. Stephens, *The Life and Letters of Walter Farquhar Hook D.D., F.R.S.* (London: Richard Bentley and Son, 1881), pp. 376, 389.

64 'Local Intelligence', *Leeds Mercury*, 9 November 1844, p. 7.

as the Vicar of Leeds, when 400 of his opponents signed a memorial denouncing his appointment. In a letter to a friend he wrote, 'they have represented me ... not only as a Papist, but as a drunkard, a gambler, &c., &c.'[65] But by 1844 his support for the Ten Hours Movement and the passing of the Leeds Vicarage Act had broadened his popular appeal and secured his legacy. It is possible that the bust entered their home at 6 Park Place, which had been designated as the vicarage for the Parish Church in 1825. If this was the case, it was not there for long. In 1845 the bust was exhibited in public twice in rapid succession. From 9 June it appeared at the Third Polytechnic Exhibition.[66] By the end of September of the same year Keyworth's bust of Hook was on display at the Second Polytechnic Exhibition at the Hull Mechanics' Institute.[67] Although it is possible that the bust travelled straight from Leeds to Hull, the Leeds Sculpture Collection points towards another possibility.

The second marble bust is unsigned and undated and it may not be possible to determine which version was made first. It is tempting to regard the inscribed version as 'the original' because it is signed and dated, its provenance is clear and the conditions under which it is displayed at Leeds Civic Hall affirm its status. But it is important to resist this designation because the construction of an opposition between original and reproduction is neither appropriate nor productive for the study of sculpture, and in some ways the unsigned version might even be regarded as superior. It does not have the cracks or accumulated dirt of the inscribed version and its surface appears softer and more sympathetic, having not been treated and polished in the same way. It is most probable that it entered the collections of Leeds Museums and Galleries via the Leeds Philosophical and Literary Society, whose building on Park Row and its contents were transferred to the city in 1921 to become Leeds City Museum.[68] Hook had been president of the Society for a brief period between 1858 and 1859, and given that busts of previous presidents and supporters of the Society decorated the Philosophical Hall, it is highly likely that the unsigned version was either donated to or purchased by the Society, possibly after Hook had left Leeds for Chichester in 1859.[69] Hook continued to act as an honorary

65 Stephens, *The Life and Letters of Walter Farquhar Hook*, pp. 204–06.

66 Hailstone, ed., *Catalogue of the Leeds Third Polytechnic Exhibition*, p. 6.

67 'Approaching Exhibition at the Hull Mechanics' Institute', *Hull Packet and East Riding Times*, 22 August 1845, p. 5.

68 Peter Brears, *Of Curiosities and Rare Things: The Story of Leeds City Museums* (Leeds: The Friends of Leeds City Museums, 1989), p. 18.

69 Edwin Kitson Clark, *The History of 100 Years of Life of the Leeds Philosophical and Literary Society* (Leeds: Jowett and Sowry, 1924), pp. 69–70.

member of the Society and left his mark on the collections, having donated a considerable number of geological and zoological specimens and the sum of £50 for new acquisitions.[70]

Hook was often called on to chair voluntary societies, meetings, exhibitions and educational causes. In addition to his campaign to establish elementary schools, he was an advocate of the Leeds School of Design, which had been established by the Leeds Mechanics' Institution in 1846. Hook was broadly in favour of state intervention in education, which placed him in conflict with the dominant and influential circle of voluntaryists who saw government interference as inherently nefarious. But among the Dissenters and nonconformists, Hook was supported by a small but prominent group of Anglicans with high social and economic status.[71]

It is no coincidence that the first teaching objects donated to the new School of Design were made by Keyworth and consisted of plaster casts of medieval woodcarvings from Lincoln Cathedral and a church at West Ardsley near Wakefield.[72] The art historian Gerardine Mulcahy noted that although Keyworth's practice was confined to portraiture and ecclesiastical monuments, his driving interests were in mediaeval sculpture, gothic architecture and the restoration of churches.[73] Keyworth went on to supply the Great Exhibition of 1851 and its second incarnation at the Crystal Palace with a substantial collection of plaster casts of medieval sculpture and architectural ornament, some of which have survived as part of Hull Museums' collections.

70 *Forty-First Report of the Council of the Leeds Philosophical and Literary Society, at the close of the session, 1860–61* (Leeds: C. Kemplay, 1861), p. 26.

71 Peter Cunningham, 'The Formation of the Schools of Design, with Special Reference to Manchester, Birmingham and Leeds' (unpublished doctoral thesis, University of Leeds, 1979), pp. 320–21.

72 'Leeds School of Design', *Leeds Mercury*, 19 June 1847, p. 5. 'The Casts presented by Mr. TW Green consist of 1,2,3,4,5,6 Casts of Poppy Heads from the Choir of Lincoln Cathedral—circa 1400 early perpendicular. 7 Casts of Poppy Head from the Church of West Ardsley near Wakefield—late perpendicular 15 Century. 8, 9, 10, 11, 12, 13 Casts of the Ellbows [*sic*] of Stalls from the Choir of Lincoln Cathedral—circa 1400'; Minutes of the Sub Committee of the Leeds School of Design (1844–54), Committee Meeting 14 June 1847, West Yorkshire Archive Service, Leeds, WYL368/23. An advert dated 11 July 1844 listed the stock of T.W. Green as 'to be disposed of', in *Bent's Monthly Literary Advertiser*, no. 482 (10 August 1844), p. 122.

73 Gerardine M. Mulcahy, 'An Eminent Sculptor: William Day Keyworth Jun. of London & Hull', *Journal of the Church Monuments Society*, XVI (2001), 126–27.

The pedestal of the inscribed bust records its transfer to Leeds Town Hall in 1859. In June of that year the Town Hall had been the setting for a lavish banquet attended by 400 people in celebration of Hook's twenty-two years in Leeds, at which he was presented with 2,000 guineas, an elaborate silver centrepiece, a clock, a candelabra and a gold casket containing £270 for Mrs Hook 'as a testimonial from the ladies of Leeds'.[74] Given the generosity of the farewell, we can interpret the donation of the bust as part of this economy of gifts transacted in public. It secured a surrogate presence for Hook in the principal civic building of the borough, which in some ways materialised the diminishing role of the Church as a new class of second- and third-generation industrialists turned to politics. On a more prosaic level, it was also an exchange that meant the Hooks were free of the burden of arranging to transport this heavy and cumbersome object almost 300 miles from Leeds to Chichester. The Town Hall Committee Minutes dated 14 September 1860 record that the bust was gratefully received and 'placed for the present, in the Mayor's apartments'. The mayor at this point was the locomotive manufacturer James Kitson (1807–85), whose second son, also James (1835–1911), would become the first Lord Mayor of Leeds in 1896.

The Picture Gallery

While the number of portrait busts increased at the 1845 exhibition, the number of painted portraits decreased.[75] Visitors entered the Picture Gallery after moving through the Hall of Sculpture, and the *Leeds Times* noted the principal paintings on display:

> A large landscape piece, by John Martin, in illustration of a well-known passage in Goldsmith's ballad of the Hermit, presents many of the characteristic beauties, and also some of the characteristic faults of that distinguished artist. The "wilds immeasurably spread" are given with the range of perspective, and vastness of effect, which few can command so well as Martin, but, at the same time, the figures are comparatively poor and insignificant. "Joash shooting the arrow of deliverance" is a very noble and striking picture. Considerable interest also attaches to a painting by Cope of Leeds, "a cup of cold water," in illustration of a passage in the 10th

[74] John Mayhall, *The Annals and History of Leeds, and Other Places in the County of York* (Leeds: Joseph Johnson, 1860), pp. 741–42.

[75] 'Leeds Polytechnic Exhibition', *Leeds Mercury*, 7 June 1845, p. 5.

> chapter of the Gospel by Matthew. The figures and the grouping are good, though the painting is not one of the most highly finished of Cope's productions... We noticed amongst the paintings by the older masters, "Vulcan forging the armour of the Gods," by Carracci, "Abraham and Ishmael," by Rembrandt, both fine pictures. These two, together with the "Youthful Shepherdess," by Gainsborough, are from the collection of Mr. Miller, of Liverpool. A small but beautiful picture, "the Crucifixion," by Correggio, which is valued at £300, is contributed by Mr. Henry Jennins, of Leeds.[76]

John Martin's epic narrative landscape paintings occupied the same critical territory as in the previous exhibition of 1843. Charles West Cope (1811–90) was also noted again, as much for his local connection as the merits of his pictures. The works attributed to Carracci, Rembrandt and Gainsborough were collected by John Miller, a merchant from Liverpool who became known as a collector and patron of the Pre-Raphaelites from the middle of the 1850s.[77] The owner of the purported Correggio, Henry Jennins, was also a merchant. In the summer of 1845 his firm of stuff and blanket merchants, Wilks, Shepard and Jennins, was reconstituted on the retirement of one of the three co-partners, John Shepard.[78] He was also the director of the Manchester, Southampton and Poole and Bradford and Wakefield railways.[79] With these commercial interests and a home on the prestigious row of Georgian townhouses in Park Place, Jennins was situated at the heart of the Mill Hill group that was so central to the culture surrounding these public exhibitions. How he came to own a Correggio worth £300—roughly equivalent to a purchasing power of £20,000 today—is more difficult to determine, as is the question of whether the attribution would withstand present standards of art historical scrutiny and provenance research.

[76] 'Leeds Polytechnic Exhibition', *Leeds Times*, 7 June 1845, p. 5.

[77] See Ann Bukantas, 'John Miller of Liverpool', in *Pre-Raphaelites: Beauty and Rebellion*, ed. by Christopher Newall and Ann Bukantas (Liverpool: National Museums Liverpool, 2016), pp. 74–78; David Taylor, *The Remarkable Lushington Family: Reformers, Pre-Raphaelites, and the Bloomsbury Group* (Lanham, MD: Lexington Books, 2020), pp. 204–05.

[78] Notice, *London Gazette*, 15 July 1845, p. 2116.

[79] Supplementary Double Sheet to *Bradshaw's Railway Gazette*, 20 September 1845, p. 607. Jennins later acted as an associate juror in the woollen and worsted category of the Great Exhibition in 1851. See *Exhibition of the Works of Industry of All Nations 1851: Reports by the Juries on the Subjects in the Thirty Classes Into which the Exhibition was Divided* (London: William Clowes and Sons, 1852), p. xxviii.

The *Leeds Intelligencer* also noted the diminished number of portraits on display, with a general increase in the number of paintings from other genres:

> The Picture Gallery next demands our attention, and it will require a very long examination indeed before the connoisseur can complete his revel amid the constellation of art which glitters around him. The pictures at the present exhibition exceed in number those which were contributed to the last. There is a preponderance of cabinet pictures over those suitable for the ordinary gallery, but each class contains many gems of the highest worth. There are not many portraits, but with historic and dramatic pieces the collection is richly supplied, and landscape and sea views are equally abundant and *recherche* [*sic*]... There are good pictures to be seen, and those who can appreciate them and have souls to feel the power of the painter's magic, will require no further incentive than this simple declaration.[80]

The *Intelligencer* reinforced its message in a more extensive appraisal published one week later. In it the reporter stressed their argument that individual taste would determine the viewer's encounter with the pictures on display: 'Each man will be for himself the best reviewer of this part of the exhibition, and therefore we leave it in the fullest confidence that the most sanguine mind will not be disappointed by actual experience.'[81]

A correspondent who wrote for the *Art-Union* journal—published in London with an international distribution—had no such reservations about articulating their thoughts about the merits and deficiencies of the exhibition and the objects within it:

> There have been many opinions as to the comparative merits of this Exhibition and the last. The present one I consider decidedly inferior in everything except sculpture, and of that we have indeed a splendid collection. Still the Exhibition is good, and one that will prove of much service to the few artists of Leeds, and which will tend, as the others have done, to act beneficially on the tastes of the people. What the London artists have sent to our Exhibition is, with one or two exceptions, decidedly bad.[82]

[80] 'Leeds Polytechnic Exhibition', *Leeds Intelligencer*, 7 June 1845, p. 5.
[81] 'Leeds Polytechnic Exhibition', *Leeds Intelligencer*, 14 June 1845, p. 5.
[82] 'Leeds', *Art-Union*, 1 August 1845, p. 267.

The writer acknowledged the contribution of the sculpture towards differentiating the 1845 exhibition from its two predecessors, although it was not enough to convince them that overall it was a better selection of objects. The wording of the article suggests that the writer was from Leeds rather than having visited from elsewhere, which was reinforced by their assessment of the work of local painters:

> And now a few words for our Leeds artists. There are one or two by Joseph Rhodes (whose pictures almost startle by their truth, nature, and pure rustic feeling) which are in every sense of the word fine.—Mr. Waller contributes a good 'Portrait of an Artist,' and also a 'Portrait of a Gentleman,' which is very good.—No. 27. 'Landscape, Morning,' by John Cruise, is a very beautiful picture, and, with one or two exceptions, the best in the Exhibition. He has also contributed a 'Sketch of Abel,' well conceived, and which with a little alteration would make a good picture.—There are some portraits by Smith, very good (a pupil he was of Mr. Joseph Rhodes, father to John Rhodes), and some good copies from Rembrandt.[83]

Since John Rhodes's death in 1842 his father Joseph Rhodes (1782–1854) had re-emerged as the primary representative of local painting. Richard Waller (1811–82) was a moderately successful painter from Skipton—a market town in North Yorkshire—who had been apprenticed to a coach maker before setting up various studios in Manchester, London and Bradford. His obituary gave a short account of how he came to settle in Leeds: 'through the influence of Mr. Kershaw, a gentleman very fond of works of art, Mr. Waller was persuaded to open a studio in Leeds. He occupied rooms in Park-square, now the site of the warehouses of Mr. Barran.'[84] While in London in 1840 he is recorded as having become acquainted with Benjamin Robert Haydon, whose ideas on public art and instruction seem to have informed his own.[85]

Leeds Art Union

For the first time in this context visitors were encouraged to become patrons of art through an Art Union established alongside the 1845

[83] 'Leeds', p. 267.

[84] 'A Yorkshire Artist's Career. The Late Richard Waller', *Supplement to the Leeds Mercury*, 29 July 1882, p. 1.

[85] 'A Yorkshire Artist's Career. The Late Richard Waller', p. 1.

exhibition. The Art Union of London had been founded in 1837 and members paid an annual subscription to be entered into a prize draw to win works of art. Each subscriber received an engraving of a contemporary painting in return for their support. The Leeds Art Union was established on a very similar model, as the *Leeds Intelligencer* explained:

> The committee have done wisely in opening an Art Union in connection with the Polytechnic Exhibition, and they will find it an excellent expedient for disposing of those "on sale" pictures which are in their care. The plan is a very perspicuous one, and at once offers to every subscriber, an equivalent for the value of his investment. The prizes are of two classes, to be competed for by subscribers of a guinea and half a guinea each. An engraving from the pictures of some of our most eminent artists will be given to every subscriber, and, even if there were not the certainty of realising this, we have not a doubt that a sufficient number of subscribers to complete the union would speedily be obtained.[86]

The art union model was presented as a low-risk means of encouraging wider audiences to own art, although the above quotation indicated that the works of art to be offered as prizes were perhaps not the most desirable or well regarded. The London Art Union was a longstanding institution and operated, with diminishing returns, until 1912. It appears to have been more common for towns and cities to set up their own art unions on a temporary basis, to sit alongside specific exhibitions. The Leeds Art Union was first advertised in the *Leeds Times* and the *Leeds Intelligencer* on 26 July 1845, explicitly 'in connection with, and for the benefit of, the Leeds Polytechnic Exhibition'.[87] The advertisement made reference to the success of similar schemes in Birmingham, Manchester and Liverpool, which connected to the same spirit of inter-city competition that pervaded the organisation of temporary public exhibitions. The committee noted that a primary objective was 'to dispose of a portion of the Works of Art intrusted [*sic*] to their Care ... and trust that they will find ample Support in carrying out a Scheme so beneficial to the Artist, and advantageous to the Public'.[88] Class No. 1 subscribers paid £1 1s. each and were given the choice between unspecified engravings by David Wilkie (1785–1841) or Thomas Lawrence

86 'The Leeds Art Union', *Leeds Intelligencer*, 2 August 1845, p. 5.

87 Advertisement and Notices, *Leeds Times*, 26 July 1845, p. 1; Advertisement and Notices, *Leeds Intelligencer*, 26 July 1845, p. 1.

88 Advertisement and Notices, p. 1.

(1769–1830).[89] These subscribers were entered into a draw to win (and choose for themselves) one of the following prizes from the works listed for sale: one valued at £20, one at £15, four at £5, five at £4 and six at £3. Class No. 2 subscribers who opted for the lower rate of 10s. 6d. could choose between engravings *after* Peter Paul Rubens (1577–1640), Bartolomé Esteban Murillo (1617–82) or Philips Wouvermans (1619–68) or Wilkie as above. The following prizes were available to these subscribers: four at £3, three at £4, two at £5 and one at £10. Tickets for subscriptions could be purchased directly from the exhibition and at the other places that sold tickets.[90] The prize draw took place at the Albion Street Music Hall at noon on 9 September 1845, three days after the official closure of the exhibition.[91] The winners were not announced in the local press and the Leeds Art Union appears to have been wound down until it re-emerged in association with the short-lived Leeds Academy of Arts from 1853.[92]

The Cabinet

The fine arts occupied a substantial part of the 1845 exhibition and their presentation continued beyond the Hall of Sculpture and the Picture Gallery into the next room: the Cabinet. As had been the case during previous exhibitions, this smaller room was primarily—although not exclusively—reserved for the display of prints and watercolours. The *Leeds Times* drew attention to two seascapes in watercolour by John Martin, 'painted on rough paper, which gives them a singular and striking effect', and portraits by the seventeenth-century court painters Peter Lely (1618–80) and Godfrey Kneller (1646–1723).[93] Like the Picture Gallery, the

89 Advertisement and Notices, p. 1. Those who purchased two shares could select one of two engravings by Wilkie: 'The Reading of the Will' or 'Blindman's Bluff'.

90 These included Messrs Slocombe and Simms on Commercial Street, Mr Henwood on Basinghall Street, Mrs Hope on North Street and the newspaper offices. Advertisement and Notices, *Leeds Intelligencer*, 16 August 1845, p. 1.

91 Advertisement and Notices, *Leeds Times*, 6 September 1845, p. 1.

92 'Leeds Academy of Arts', *Leeds Times*, 24 December 1853, p. 5.

93 'Leeds Polytechnic Exhibition', *Leeds Times*, 7 June 1845, p. 5. The two watercolours by Martin were listed as No. 36, *The Open Sea*, and No. 46, *Sappho*, loaned by the artist and both for sale. The work by Lely was No. 14, *Portrait of the Countess of Richmond*, loaned by M.J. Rhodes, and by Kneller, No. 66, *Portrait of a Lady*, contributed by Thomas Brown, Esq. See Hailstone, ed., *Catalogue of the Leeds Third Polytechnic Exhibition*, p. 24.

Cabinet also contained a large number of 'curiosities in glass cases under the broad classification of "British and Foreign Antiquities, &c."'

The *Leeds Times* noted just a fraction of the diverse range of objects on display:

> Amongst other articles of curiosity, we were shown a remarkable chess and backgammon board, the black draughtsmen belonging to it being carved with the heads of the Cæsars, and the white ones with the heads of the German Emperors; a large and ancient wheel-lock pistol, elaborately inlaid with ivory, and supposed to be a century and a half old; specimens of Richardson's rubbings of monumental brasses; specimens of wood carving intended for competition for the adornment of the new Houses of Parliament; a design for a stained glass window representing, by their armorial bearings, the successions and alliances of the sovereigns of England from the Conquest down to the present period; a quantity of robes and arms from New Zealand, and a few articles from China.[94]

These 'few articles from China' were displayed under the category of 'Chinese Costumes, &c, taken during the late war'.[95] It is likely that these objects were looted during the First Opium War between Britain and the Qing dynasty, which was fought between 1839 and 1842. Eighteen objects were loaned by five men, and they included textiles, edged weapons and a firearm. The *Leeds Times* commented: 'Some very curious Chinese dresses and armour, many of them taken during the late war by Lieutenant Cator, have been contributed by the Rev. C. Cator, Mr. Robert Barr, and Mr. Dickinson, of Headingley. There are also several specimens of carving, scales and instruments of torture from China, which will be looked at with interest.'[96] Lieutenant Cator was Sir John Bertie Cator (1820–87), promoted to that rank in 1842 after serving in the British Navy on HMS *Herald* where 'he beheld the capture of several forts, and of the last one protective of the approaches to Canton'.[97] The lender of the objects was either his elder brother or his father: Revd Charles Cator (1812–48) or Revd Charles Cator (1786–1872). The family had relatively local connections in Skelbrooke, a village to the north-west of

94 'Leeds Polytechnic Exhibition', *Leeds Times*, 7 June 1845, p. 5.

95 Hailstone, ed., *Catalogue of the Leeds Third Polytechnic Exhibition*, p. 34.

96 'The Leeds Polytechnic Exhibition', *Leeds Times*, 14 June 1845, p. 5.

97 William R. O'Byrne, *A Naval Biographical Dictionary: Comprising the Life and Services of Every Living Officer in Her Majesty's Navy, from the Rank of Admiral of the Fleet to that of Lieutenant, Inclusive. Compiled from Authentic and Family Documents*, vol. 1 (London: John Murray, 1849), p. 180.

Doncaster in South Yorkshire, and Stokesley, a market town in North Yorkshire where Charles Cator senior was rector at the Parish Church.[98] Since the 1841 Census placed Charles Cator junior in Middlesex in that year and the place of his death and burial in 1848 was Wellington in New Zealand, it is perhaps more likely that the lender to the 1845 exhibition was his father.[99] Cator contributed six of the eighteen items: a white silk scarf, a lady's pocket, gentlemen's quilted stockings, a blue silk dress and two hats.[100] Cator lent a further twenty-four objects, the vast majority of Chinese origin, that were also displayed in the Cabinet, though without the specific designation of having been 'taken during the late war'.[101] These items sound innocuous enough until we consider what the conditions might have been under which they were stolen from their Chinese owners, alive or dead. Aside from the acknowledgement of their origin in the catalogue and the local periodical press, the exhibition appeared neither to celebrate nor condemn the way in which they came to be on display in Leeds. That these objects were made available for public consumption does, however, indicate that the committee were complicit in the ideological messaging that tacitly communicated the perceived legitimacy of both the military action in China and the practice of stealing and displacing the material culture of other countries.

The *Leeds Intelligencer* interpreted the contents of the Cabinet as a *wunderkammer*, with the only organising principles being the extent to which the objects stimulated curiosity in the viewer and communicated the taste and learning of the collector/curator:

> Leaving the Picture Gallery the visitor finds himself within the Cabinet, which is decorated with an assortment of delicate and valuable rarities. We have an abundance of artificial curiosities, many of which are the contributions of foreign climes. There is a profusion of Chinese articles of dress and accoutrements, which form not the least attractive portion of the display. The character of the whole is in strict keeping with the title of the apartment. It

[98] Census Returns of England and Wales, 1851, Class: HO107; Piece: 2362; Folio: 91; Page: 7; GSU roll: 87635-87636, and Class: HO107; Piece: 2376; Folio: 91; Page: 2; GSU roll: 87665.

[99] Census Returns of England and Wales, 1841, Class: HO107; Piece: 679; Book: 4; Civil Parish: St Marylebone; County: Middlesex; Enumeration District: 6; Folio: 35; Page: 22; Line: 13; GSU roll: 438795. See also New Zealand Cemetery Records, 1800–2007, Anglican Burials, Wellington/Hutt Area, p. 4.

[100] Nos. 1–5 and 10. See Hailstone, ed., *Catalogue of the Leeds Third Polytechnic Exhibition*, p. 34.

[101] Hailstone, ed., *Catalogue of the Leeds Third Polytechnic Exhibition*, p. 34.

> is a repository of matters of taste and vertu, and is an assemblage of many articles, the glitter of which gives a pleasant variety to the grandeur of the spectacle through which we have previously passed. It presents a brilliant array of many gaudy prettinesses ... and at this stage affords a pleasing and grateful deviation from the sublimer works which have before engaged the attention.[102]

The domestic scale of the Cabinet also admitted the work of women, to a very limited extent. The *Leeds Times* reported: 'We noticed some handsome embroidery by a lady of Leeds in the cabinet, and were told that two specimens of ornamental turning in ivory, are contributed by the same fair artist.'[103] Unfortunately both the artist and the contributor were unnamed in the catalogue.[104] Of the paintings in the Cabinet, three were by women and two were contributed by women.[105]

The Hall of Caves

Once visitors had encountered these 'foreign' objects in the Cabinet they entered a liminal space on their journey to the Saloon. The Hall of Caves was a new attraction, described as 'a narrow winding and over-arched avenue, with caverns on either side, containing representations of scenes in many lands'.[106] The *Leeds Intelligencer* provided a more detailed account of the arrangement of the space and its contents:

> A very singular, and magician-like apartment, or rather series of apartments, which will doubtless, as it did at the recent Liverpool Exhibition, form a source of great attraction and pleasing wonder to all who enter its labyrinthian recesses. It is erected in a building to the rear of the Music Hall, but there is a covered passage between

[102] 'Leeds Polytechnic Exhibition', *Leeds Intelligencer*, 14 June 1845, p. 5.

[103] 'The Leeds Polytechnic Exhibition', *Leeds Times*, 14 June 1845, p. 5.

[104] Case D, No. 1: 'English Scarf, richly embroidered in Floss Silk, by a Lady', Hailstone, ed., *Catalogue of the Leeds Third Polytechnic Exhibition*, p. 34.

[105] Painted by women: No. 26, *Madonna* by Elizabeth [Elisabetta] Sirani, 'a favorite [*sic*] pupil of Guido', contributed by Dr Hopper, p. 23; No. 32, *Landscape* by Charlotte Nasmyth, contributed by Dr Hopper, p. 24; No. 38, *Gipsey* [*sic*] *and Child* by Miss Chappell, contributed by the artist, p. 24. Contributed by women: No. 13, *Fish* by A. Breil, contributed by Mrs Carrett, p. 23.

[106] 'Leeds Polytechnic Exhibition', *Leeds Times*, 7 June 1845, p. 5.

> the two, so that the visitor has no difficulty in passing from the one to the other: indeed, the whole seems but one suite of rooms.[107]

The *Intelligencer* revisited its account one week later and added:

> when the excursionist has threaded his way along the obscure path which leads to the recess, he comes in succession upon several illuminated vaults which are hung with scenic drapery that gives fantastic representations of romantic views in a pleasant land. One of these is in imitation of the interior of a cavern of spar, and, being bespangled with corruscant [*sic*] particles, has a very pleasing and appropriate effect.[108]

This staging of landscapes and geological features was connected with the proliferation of panoramas during this period. The *Journal of the Leeds Polytechnic Exhibition* noted the divided reception of this new attraction through the correspondence received from those that had visited:

> Letter 3 begs most humbly to ask the Curators what is the meaning and design or that burlesque, the Hall of Caves; is glad that we have lately closed up a considerable part of the openings, and begs to submit to our wisdom, whether it would not be a very great improvement to shut up the views altogether. While Miss Amelia Lucy Caudle, in a most exquisitely scented "hot pressed satin-note," Letter 4, begs us whatever we do else, not to disturb that enchanting place the Hall of Caves.[109]

Unhappily for the first correspondent, the 'burlesques' continued beyond the Hall of Caves. Before the visitor entered the Saloon they encountered Mr Austin's 'Happy Family', which consisted of 'a most extraordinary collection of beasts and birds, which, notwithstanding their natural hostility to each other in a state of nature, live here together in one cage in perfect harmony, the hawk feeding the starling, the cat suckling the rat, and all sharing bed and board together'.[110] John Austin toured his menagerie around London and across the country

[107] 'Leeds Polytechnic Exhibition', *Leeds Intelligencer*, 7 June 1845, p. 5.

[108] 'Leeds Polytechnic Exhibition', *Leeds Intelligencer*, 14 June 1845, p. 5.

[109] 'Editor's Box', *Journal of the Leeds Polytechnic Exhibition, for providing Public Walks and Baths*, No. 5 (7 July 1845), p. 34.

[110] 'Leeds Polytechnic Exhibition', *Leeds Times*, 7 June 1845, p. 5.

before it was purchased by the famed American showman P.T. Barnum (1810–91) while on display in Coventry in 1844. Barnum recalled, 'we visited an exhibition called the "Happy Family," consisting of about two hundred birds and animals of opposite natures and propensities, all living in harmony together in one cage. This exhibition was so remarkable that I bought it for $2500 (£500), and hired the proprietor to accompany it to New-York, where it has ever since been an attractive feature in my Museum.'[111] It was not Austin who accompanied Barnum but John Sutton, Austin's stepson, with whom he had worked. Sutton took only a selection of animals to New York, and presumably it was from the remaining collection that Austin made a display in Leeds in 1845, as it was reported that Austin had amassed some 700 animals over twenty years.[112] Like the Hall of Caves, the 'Happy Family' operated at the intersection of science and the sideshow and was omitted from the catalogue. Some contemporary commentators interpreted the display as a wholesome demonstration of the possibility of overcoming difference to live together in peace. An article in the *Student* journal asked: 'Is it not too much to believe that many a person who gave a penny to look at Mr. Austin's Happy Family may have had his mind awakened to the effects of habit, and the influence of gentle discipline and kindness.'[113] These animals were almost certainly not treated with 'gentle discipline and kindness', but it suited the Christian narrative to encourage comparable obedience and restraint in human behaviour.

There was one final attraction, or distraction, before the visitor proceeded. In a small room between the Hall of Caves and the Saloon, 'Mr. Walker, from Liverpool' was available to cut silhouette portraits as souvenirs.[114] This was not the only means of capturing likenesses on offer at the exhibition: the nearby Photographic Gallery had been engaged to offer portraits at half the usual cost.[115] The full price was one guinea for a photographic portrait 'in a neat Morocco Case', so the exhibition price would have been 10s. 6d.—still a considerable sum,

[111] P.T. Barnum, *Struggles and Triumphs: Or, Forty Years' Recollections of P.T. Barnum* (London: Samson Low, Son & Co., 1855), p. 280.

[112] N.A. Calkins, ed., 'Happy Family', *Student: Family Miscellany and Monthly School-Reader, devoted to the physical, moral and intellectual improvement of youth; embracing the natural sciences, natural history, biography, history, and music*, vols. VII and VIII (New York: N.A. Calkins, 1854), pp. 116–17.

[113] Calkins, ed., 'Happy Family', p. 116.

[114] 'Leeds Polytechnic Exhibition', *Leeds Times*, 7 June 1845, p. 5.

[115] 'Leeds Polytechnic Exhibition, for providing Public Walks and Baths', *Leeds Mercury*, 31 May 1845, p. 4.

beyond the reach of most visitors.[116] More than a simple commercial transaction, it was intended to display and explain a photographic process that remained novel.[117] The Photographic Portrait Gallery opened at 27 Park Row on 11 April 1842, having secured an exclusive licence to produce daguerreotypes in Leeds.[118]

The Saloon

Finally, visitors entered the Saloon—the largest single space and once again the site of some of the most dynamic objects. The *Leeds Times* published an extended description of the Saloon with its varied contents:

> The orchestra has been fitted up as a proscenium, surmounted by one of the Westminster Cartoons representing the "Throne of Intellect" by Thomas, and inscribed "Investigation calm whose silent powers command the world." Between two pillars, and in front of the proscenium, falls a large screen on which will be exhibited the dissolving views, the microscope, and that beautiful new optical instrument the Chromatrope, which is now exciting so much attention in the London Polytechnic Institution. At the foot of the orchestra is that splendid scientific instrument, the Hydro-electric machine, invented by Armstrong. With this instrument all the most imposing electrical experiments will be exhibited on a gigantic scale. On one of the side walls of the saloon are displayed diagrams illustrative of the principle of three different working models of the atmospheric railway; and underneath them are the three model atmospheric railways themselves, which will be put in operation at specific periods, and lectures will be delivered respecting them two or three times a day. Further on, on the same side, are several models or smoke burning apparatus. On the opposite side are tables covered with models of machinery, and above these, on the wall, is one of the Westminster Cartoons, representing a council of ancient Britons. Close to the wall under the gallery is a museum of natural curiosities. Here also has been erected a circular railway, round which a model galvanic carriage will travel at intervals.[119]

116 Advertisements and Notices, *Leeds Mercury*, 8 June 1844, p. 1.
117 'Polytechnic Exhibition', *Leeds Times*, 31 May 1845, p. 4.
118 Advertisements and Notices, *Leeds Mercury*, 15 May 1847, p. 1.
119 'Leeds Polytechnic Exhibition', *Leeds Times*, 7 June 1845, p. 5.

The Saloon also attracted both praise and derision, with some asking 'why we have left the Saloon so bare,—no fountain, no canal, no steam ships', and others celebrating their absence: 'banishing those toys that disgraced the Saloon on the last occasion, and making better accommodation for the visitors to the Concerts'.[120] The visitor's experience of the Saloon was contingent on the time of day at which they visited; it was the space that was most subject to change, based on scheduled experiments, lectures and musical performances. These performances extended beyond the official closure of the exhibition, as they had in 1843. On 6 September 1845 the *Leeds Times* reported that 'The Exhibition will definitely close this day, but during the period requisite for dismantling the rooms, the committee, we understand, are disposed to continue the musical entertainments in the Saloon; they are now in negotiation with the Distin family for this purpose.'[121] The Distin family were quickly engaged for four performances over the course of September 1845, advertised in the *Leeds Mercury* in the following terms: 'Mr. Distin and his four sons, The Original and only Performers on the newly invented instruments, the Sax Horns, whose Performances have created such an extraordinary sensation on the Continent and in London.'[122] The *Leeds Times* published an effusive review after the first performance:

> Nothing could exceed the power, brilliancy, and sweetness of execution, which distinguished their masterly performance on those unique and melodious instruments, the Sax Horns. They were listened to throughout with evident demonstrations of delight, and repeatedly encored. The numbers assembled on Wednesday, in spite of the great drawback of insufficient advertisement, were enough to give a presage of the brilliant support which these peculiar and delightful concerts are likely to obtain.[123]

Although entirely positive in its assessment of the concert, the review does suggest that the haste with which they were arranged had prevented the usual amount of notice being given. The Distin family were particularly well suited to perform in the context of a polytechnic exhibition, as their instrument of choice—the saxhorn—was invented

[120] 'Editor's Box', *Journal of the Leeds Polytechnic Exhibition, for providing Public Walks and Baths*, No. 5 (7 July 1845), p. 34.

[121] 'The Exhibition', *Leeds Times*, 6 September 1845, p. 4.

[122] Advertisements and Notices, *Leeds Mercury*, 13 September 1845, p. 5.

[123] 'The Distin Family', *Leeds Times*, 13 September 1845, p. 4.

towards the end of the 1830s and patented by Adolphe Sax in 1845, making it as novel as the other emerging technologies on display in the Saloon.

That the Third Polytechnic Exhibition proved to be the final iteration of this series suggests that it suffered from the law of diminishing returns, not only in financial terms but also in the number of visitors and the extent of public interest more broadly. The *Art-Union* published a short review three weeks after its closure, with a focus on the extent to which it was of benefit to living artists. The author lamented: 'It is to be regretted that the contributions of artists were very few in number—not more than sixty-four altogether, including drawings, and four of these were of high character.'[124] Taken as a fraction of the thousands of objects on display, this number does appear to be small, but taken on its own terms it is not insignificant for a regional town to have brought together works by this number of living artists. The article concluded on a more optimistic note: 'We rejoice to learn there is a great probability of an annual exhibition being established in this wealthy manufacturing town. It is, indeed, a reproach to Leeds to have been so far, and so long, behind Manchester, Liverpool and Birmingham.'[125] This statement must have stung as it cut right to the heart of the problem; a problem as much of perception as actual activity. While the 1845 exhibition may have superficially resembled its predecessors of 1839 and 1843, it was distinctive and important for its display of contemporary British sculpture in particular. It doubled down on the presentation of both the fine arts and the fairground, with exhibits such as the Hall of Caves and Austin's 'Happy Family' having been toured, viewed and interpreted in a range of public contexts, none of which were ostensibly in service of the cause of public health and sanitation.

For public walks and baths?

If the success of the exhibition is judged on its material contribution to the public walks and baths that its projected profits were intended to generate, then it is difficult not to conclude that it failed in its most fundamental object. Baths and outdoor pleasure grounds already existed in the local area, although their entrance fees were prohibitively expensive for the working classes. The Leeds Public Baths Company had been established in 1826 and the Leeds Zoological and Botanical

124 'Leeds', *Art-Union*, vol. 7, 1 October 1845, p. 317.
125 'Leeds', p. 317.

Gardens had been open to the public since 1840.[126] As free amenities offering fresh air and exercise beyond the polluted urban centre, Woodhouse Moor was not acquired by Leeds Corporation until 1857 and Roundhay Park opened to the public later still in 1872. After the Baths and Washhouses Act of 1846 a committee was formed to consider an application for Leeds, but as Brian Barber noted in his history of Leeds Corporation, even though a Member of Parliament for the borough offered a site in 1847, the idea was not realised in this context until 1878.[127]

As the exhibition of 1845 was not organised under the auspices of the Leeds Mechanics' Institution and Literary Society, it has been more difficult to track its financial results in the absence of annual reports and statements in the local press. The absence of congratulatory public commentary is in itself significant, suggesting that there was little to celebrate. The presumption that the exhibition was a failure in monetary terms is supported by articles written the following year, when questions began to be asked about the future of the public walks and baths to which the profits should have been dedicated. These discussions were bound together with the future of the Leeds Zoological and Botanical Gardens, which had never found a secure financial footing. A meeting of the shareholders was held in May 1846 where it was reported that 'there is not the slightest prospect of the gardens being kept open under the present management beyond, at the latest, the end of October [1846]'.[128] The gardens had been propped up for the previous three years by a number of private individuals who paid the interest on the debt of £4,500. Expenditure continued to exceed income and, as a result, J.G. Marshall suggested that the Town Council purchase the gardens for public use 'as a source of healthful and elevating amusement'.[129] The rationale for this proposed action once again referenced the lack of amenities in Leeds, and how the town compared unfavourably with its peers:

> Now that public parks and walks are being established in many of the large towns of England, a town of such importance as Leeds

[126] R.J. Morris, 'Clubs, Societies and Associations', in *The Cambridge Social History of Britain 1750–1950*, ed. by F.M.L. Thompson, vol. 3 (Cambridge: Cambridge University Press, 1990), pp. 395–444 (p. 412).

[127] Brian John Barber, 'Leeds Corporation, 1835–1905: A History of its Environmental, Social and Administrative Services' (unpublished doctoral thesis, University of Leeds, 1975), pp. 148–49, 308. The MP in question was either William Beckett or James Garth Marshall.

[128] 'Leeds Zoological and Botanical Gardens', *Leeds Times*, 2 May 1846, p. 5.

[129] 'Leeds Zoological and Botanical Gardens', p. 5.

> ought not, in this respect, to occupy an isolated position. We should scarcely think that for such an object as placing these gardens in the hands of the public a trifling rate would be objected to, especially if a grant could also be obtained from the government. The gardens once purchased, we have no doubt they might be made to pay expenses, and perhaps yield a revenue towards public baths, &c.[130]

There was no mention of the application of profits from the Third Polytechnic Exhibition for this object, which suggests that they were too small to be of any use; that there were not any at all; or worst of all, that it had operated at a loss.

One week later the *Leeds Intelligencer* took up the story and berated the town for its inability to translate discussion of improvements to public health into concrete action:

> The inhabitants of this rich and populous manufacturing town are, in connexion with this subject, on the point of being signally disgraced. The Polytechnic Exhibition, got up last year for the establishment of baths and public walks, was, as the gentlemen of the guarantee fund may perhaps remember, a most disastrous failure. The Sanatory [*sic*] Committee too, partly aiming at the same object, after collecting some valuable evidence on the wretchedly neglected state of the town, and the absolute necessity of some searching measures of relief, met with so little encouragement, that it seems to have died away from pure inanition and neglect.[131]

This article was the first to break ranks and name the 1845 exhibition as not simply unsuccessful, but a 'disastrous failure'. The *Leeds Intelligencer*—alongside the *Leeds Mercury* and *Leeds Times*—had been unwaveringly loyal in their support and continued to promote the exhibitions in the strongest terms until the end, but it must have been clear that public interest had diminished and that they would not achieve anything like the success of the First Public Exhibition of 1839.

The article did not end there, however, warning that 'what follows is even worse'.[132] What followed was a scathing account of the position of the Leeds Zoological and Botanical Gardens:

[130] 'Leeds Zoological and Botanical Gardens', p. 5.

[131] 'The Intelligencer', *Leeds Intelligencer*, 9 May 1846, p. 4.

[132] 'The Intelligencer', p. 4.

> Not only are we to go forth into the world as people too faint-hearted to embark in [*sic*] new enterprizes [*sic*] of elegant utility and social improvement, but we are even to be under the ridiculous reproach of being too supine or too stingy to support those in which were have embarked. Our beautiful Botanical Gardens are on the point of being broken up. As one of the Shareholders manfully expressed it "Leeds has not spirit enough" to preserve them. We earnestly entreat it to rescue itself from the mortifying imputation.[133]

The author supported Marshall's proposal to take the gardens into public ownership, and more specifically, to allow the working classes access: 'Instead of absurdly excluding them from the gardens on the only day they can possibly enjoy them, let that day be especially devoted to them.'[134] The day in question was Sunday and the argument was a familiar one: that the values and virtues represented by the observation of the Sabbath were best encouraged not by preventing secular public activities, but by the management of behaviour through the opening up of spaces 'where they may become familiar with elegance and order, and have the means of developing those tastes and habits which are the sources of cheerful enjoyment to every class of society, to none more than the poorest'.[135]

It is clear that spaces for walking and bathing already existed when the idea was formed to dedicate the profits of the proposed Third Polytechnic Exhibition to this cause. What did not yet exist were comparable spaces for walking and bathing that were economically and socially accessible to those unable to pay for the privilege. It seems not to have occurred to the organisers to attempt to municipalise existing amenities—even those that were failing as commercial concerns—rather their intention was to maintain separate social spheres between the classes. The cause on its own terms seems not to have generated sufficient support, coupled with the expense of staging a spectacle that could compete with, if not exceed, what audiences had come to expect from the previous two exhibitions.

133 'The Intelligencer', p. 4.
134 'The Intelligencer', p. 4.
135 'The Intelligencer', p. 4.

Conclusion

The 1843 exhibition had a different inflection from its two predecessors. Its stated aim to provide public walks and baths presented a more direct social purpose than supporting the Leeds Mechanics' Institution and Literary Society. Although education remained fundamental, the organisers shifted their priorities to public health and social reform in the wake of new research into the poor sanitary condition of the town. The governance of the exhibition altered too, from the committees formed through the subscriber democracies of the local voluntary societies, to four trustees with roles in local and national government and the Church. Civic rivalry continued, and this time the organisers looked to Manchester, Birmingham and Sunderland as towns in pursuit of their own public walks and baths through both parliamentary grants and local subscriptions. Leeds was also positioned in competition with Liverpool, Manchester and London to encourage the support of its local population.

As a result of the slightly extended time allocated to organising the exhibition, it was possible to secure prestigious loans from public and private collections in London, the largest of which had been generated by the competitions to design works of art for the rebuilt Houses of Parliament. The statuary intended for this purpose formed the core of a new Hall of Sculpture, which replaced the previous Museum of Antiquities through which the visitor entered the exhibition. The display of contemporary sculpture gave the exhibition a distinctive character and connected Leeds with the site of national governance, to which it had only had the right to send two elected Members of Parliament to represent the borough since 1832.

While the cultural heritage of countries colonised by or in conflict with Britain had been displayed in previous exhibitions, at the 1843 exhibition the origin of these objects was made more explicit. The catalogue noted that examples of Chinese textiles, arms and armour had been taken during the First Opium War of 1839–42. The organisers of the 1843 exhibition used this public platform to attempt to elevate the town's local and national status, but also to position Leeds as part of the wider British Empire.

The working classes remained a primary audience of the exhibition, reflected in the maintenance of the cost of a day ticket at 6d. The doubling of the cost of a season ticket, however, resulted in a substantial public backlash that exposed the class divisions that public exhibitions were intended to at least partially obscure. The organisers did eventually capitulate, but visitor figures did not substantially recover, and the

Leeds Polytechnic Exhibition of 1845 did not ultimately contribute to the provision of public walks and baths in the town. We might wonder why the organisers thought that the event would generate more money than the 1843 exhibition, when the 1843 exhibition had already suffered from diminishing returns relative to the 1839 exhibition. The profits had an inverse relationship to the confidence projected by successive organisers, whose belief in the social, cultural, economic and political utility of their collective endeavour only increased over time. Those behind the exhibitions almost enacted the symptoms of a gambling addiction, which distorted the likelihood of the next move resulting in a win that would relieve previous losses. However, although Leeds had a habit of throwing good money after bad through its public exhibitions, financial success and audience attendance are not the primary metrics by which this book judges their significance.

4

Exhibitions and Art Education in Leeds, 1846–56

Following three polytechnic public exhibitions staged at the Albion Street Music Hall between 1839 and 1845, the locus of exhibition making in Leeds shifted in the middle years of the nineteenth century. The profits from the 1839 exhibition had secured premises for the Leeds Mechanics' Institution and Literary Society at 12 South Parade, which allowed the Society to provide an expanded educational programme and associated exhibitions. Although the diminished return on the 1843 exhibition had not liberated the Society from its remaining liabilities, its relatively stable position provided the opportunity to extend its role through the campaign to establish a Government School of Design in the town. Approved in the winter of 1846, the Leeds School of Design opened access to both a centralised repository of objects for display and an emerging national network of branch schools, whose local governing bodies were also engaged in an experiment to connect the formation of teaching collections with the provision of public exhibitions, local museums and an attendant improvement in industrial manufactures associated with the design reform movement in Britain. This chapter bookends two such exhibitions: the inaugural Exhibition of French Manufactures in 1846–47 and the local presentation of the Circulating Collection or 'Travelling Museum' almost a decade later. Between these two exhibitions, staged as part of a nascent nationalised system of art and design education, a group of artists attempted to establish a regional outpost of the Royal Academy of Arts in Leeds. This short-lived attempt to support training in fine art through seasonal selling exhibitions reveals the different priorities of local artists, industrialists and the intelligentsia as they sought to define and advance the cultural life of the town.

The Exhibition of French Manufactures at the Leeds School of Design, 1846–47

On Saturday 5 December 1846 'a brilliant and unique collection of French manufactures', purchased by the British government for exhibition at the Schools of Design, was displayed to the public at the Leeds Mechanics' Institution and Literary Society at 12 South Parade. The collection was presented alongside the recently granted teaching collection of plaster casts for the new Leeds School of Design, which was shortly to open on 11 January 1847.[1] The collection of French manufactures was sent to new regional Schools of Design as a material manifesto, intended to establish the ethos and secure the influence of the Head School at Somerset House in London. As part of his role as director of the schools, the artist Charles Heath Wilson (1809–82) had been sent to Paris to purchase objects from the Exposition de l'industrie française, année 1844.[2] The objects he acquired were intended to augment the teaching collections granted to the schools by providing examples of contemporary continental manufactures. The council of the Schools of Design was keen to improve the quality and relevance of the objects circulated to the schools and Wilson was authorised to spend up to £1,400 in Paris for this purpose.[3] The council later retracted its initial plan to distribute objects for permanent use in the schools in favour of the formation of a touring collection that would be presented as a temporary public exhibition at each school. The objects were to travel by rail in purpose-built cases accompanied by a descriptive catalogue.[4] In the absence of a catalogue issued by the Leeds School

1 'Leeds School of Design', *Leeds Mercury*, 12 December 1846, p. 8. The plaster casts exhibited alongside the French manufactures were recorded in the same article: 'The statues, size of life, are the Apollo Belvidero [*sic*], the Venus di Medici, Discobolus, Germanicus, Venus Milos [*sic*], and Fighting Gladiator, with busts of Niobe, Titan Antinous, &c.; masks of Apollo, Venus, Juno, Hercules, and Jupiter; numerous anatomical and other specimens of hands, feet, arms, &c.; pilasters from the tomb of Louis XII, large specimens of Trajan Frieze, Roman arabesque and cornice; copies of antique statues, by machinery, and a considerable number of architectural specimens.'

2 Council of the School of Design, *Third Report of the Council of the School of Design, for the year 1843–44* (London: printed by William Clowes and Sons for Her Majesty's Stationery Office, 1844), pp. 31–32.

3 Council of the School of Design, *Fourth Report of the Council of the School of Design, for the year 1844–45* (London: printed by William Clowes and Sons for Her Majesty's Stationery Office, 1845), p. 11.

4 Council of the School of Design, *Fourth Report of the Council of the School of Design*, p. 21.

of Design, the objects exhibited in Leeds as part of this scheme can be partially recovered through *The Fourth Report of the Council of the School of Design* and a descriptive catalogue compiled by the Norwich School of Design, published in 1846.[5] The introduction to the section on contemporary continental objects invoked both admiration and anxiety when it described the 'extraordinary and rapid progress which the French are making in a number of important branches of trade' and their excellence in bronze casting, ceramics, silks, carpet and damask weaving. Earthenware was positioned as a field in which 'the French are advancing', which was a phrase that seems to have been calculated to cause concern and motivate British manufacturers to exceed their great rivals across the Channel.[6]

As Clive Wainwright identified through his research into the early collections of the Government School of Design, many of these objects were retained and became part of what is now the permanent collection of the Victorian and Albert Museum. Wainwright considered them to represent the foundation of the collection of contemporary decorative arts.[7] In Leeds, the production of and trade in decorative commodities informed by French designs was already firmly established. The suppliers and manufacturers of Parisian paper hangings, Trumble and Cooke, were particularly notable for their support of the Leeds School of Design and took on a former pupil as an assistant designer in 1848.[8]

The objects Wilson purchased from the Paris Exposition in 1844 for exhibition in the British Schools of Design were not intended for mass production or consumption. He selected exceptional and elaborate examples, often produced specifically as exhibition pieces. For example,

5 Council of the School of Design, *Fourth Report of the Council of the School of Design*, pp. 38–39; Norwich School of Design, *A Catalogue of the Casts, Figures, &c. Furnished by Government, for the use of the Students at the School of Design at Norwich, and also of a variety of Articles of French Manufacture, purchased by Government at the Great National Exhibition at Paris in 1844, which have been deposited at the School of Design, Norwich for One Month* (Norwich: printed by Charles Muskett, 1846).

6 Norwich School of Design, *A Catalogue of the Casts, Figures, &c.*, p. 7.

7 Clive Wainwright, 'The Making of the South Kensington Museum I: The Government Schools of Design and the Founding Collection, 1837–51', *Journal of the History of Collections*, 14 (2002), 3–23 (p. 19). Wainwright also detailed the objects that have not survived or been identified: 'None of the embroidered silks, velvets, chintzes, tapestries, carpets, wallpapers and stamped leather ever made it into the 1868 *Inventory*, and nothing in these categories has survived with 1844 numbers.' See Wainwright, 'The Making of the South Kensington Museum I', p. 7.

8 'Leeds School of Design', *Leeds Mercury*, 12 August 1848, p. 7.

article number 60 in the catalogue issued by the Norwich School of Design was listed as 'Blue Vase, from Sèvres, called the *Adelaide Vase*' and purchased for £12 11s. 2d.[9] Although a contemporary work, the vase was decorated in a sixteenth-century French Renaissance style, for which Wilson was known to have a particular penchant.[10] The account of the item provided in the catalogue assumed a high degree of pre-existing knowledge of the source material: 'This beautiful article is painted in imitation of the celebrated old enamels of Limoges. The manufacture at Sèvres is too well known to need any account of it; but the artist-like skill with which the painting on the vase is executed, is worthy of notice.'[11] The use of the descriptor 'artist-like' was particularly significant, as the students of the Schools of Design were not intended to aspire to the status of artists, although they were expected to mimic their accomplishments. Similarly, article number 67, a silver vase produced by the atelier of Frédéric-Jules Rudolphi, was displayed as a unique showpiece, a model of the collaborative hierarchy between the designer and maker. At £40 it was one of the most expensive purchases made by Wilson in Paris.[12] It was the unnamed chasers who were particularly recognised, however, for their contribution to the success of the piece. The elevated status of these artisans was used to legitimise the project of the British Schools of Design; the education, knowledge and high wages of the French artisans were presented as an aspirational for their British counterparts.[13] A statuette of *Icarus* by Philippe Grass, cast in bronze by the founders Eck and Durand, was also purchased for the sum of £10 8s.[14] The descriptive catalogue did not discuss the formal merits of this piece as a sculpture, but instead used it to illustrate

[9] Norwich School of Design, *A Catalogue of the Casts, Figures, &c.*, p. 8.

[10] Stuart Macdonald, *The History and Philosophy of Art Education* (London: University of London Press, 1970), p. 89.

[11] Norwich School of Design, *A Catalogue of the Casts, Figures, &c.*, p. 8.

[12] Science and Art Department of the Committee of Council on Education, *Inventory of the Objects in the Art-Division of the Museum at South Kensington, Arranged According to the Dates of their Acquisition. Vol. I. For the Years 1852 to the End of 1867* (London: printed by George E. Eyre and William Spottiswoode for Her Majesty's Stationery Office, 1868), p. 2. The object was recorded as: 'Bottle. Oxydised silver. Bulbiform, chased with foliage; two cupids form the handles. *French*, modern. (Rudolphi, Paris.) H. 9½ in., diam. 4½ in. Bought, 40l. 919.—'44.'

[13] Norwich School of Design, *A Catalogue of the Casts, Figures, &c.*, pp. 9–10.

[14] Science and Art Department, *Inventory of the Objects in the Art-Division*, p. 40. The object was recorded as: 'Statuette. Bronze. Icarus. *French*, modern. (By P. Grass, cast by Eck and Durand.) H. 21½ in., W. 10 by 8 in. Bought (French Exhibition 1844), 10l. 8s. 379.—'44.'

another perceived virtue of continental modes of production: 'the French sculptor has a great advantage in respect to the facility with which he may get his works cast in bronze, and the moderate price of the operation'.[15] The manufacturing process was foregrounded to draw particular attention to the infrastructure required to operate on the same terms as the French.

After these objects had been displayed in Leeds for one month Wilson requested that the collection be sent on to Stoke-on-Trent, where a School of Design had been established in 1846 to serve the potteries district.[16] The collection of French manufactures was exhibited in Leeds for a second time on 2 February 1847.[17] The committee of the Leeds Mechanics' Institution and Literary Society reported that the attendance had been high, with manufacturers having travelled from nearby Bradford and Halifax to see the exhibition.[18] The *Leeds Mercury* provided a favourable review, which noted that the room provided for the exhibition had been repainted 'for the purpose of "throwing up" the statues and models placed near the walls', and called for a larger space to be allocated to ease overcrowding.[19]

The council of the Schools of Design was also satisfied with the consequences of having circulated the collection around the fledgling branch schools outside London. It reported that the objects had been received with interest by the intended audience of students and manufacturers, who in some instances took the opportunity to produce copies of the works they had seen.[20] This positive assessment of the

[15] Norwich School of Design, *A Catalogue of the Casts, Figures, &c.*, p. 10.

[16] 'Resolved that the specimens of French Manufacture be sent to Stoke upon Trent as requested by C.H. Wilson Esq. Director of the Schools of Design in his letter dated 18th Jan. – and that the thanks of the Committee be presented to the Council of the Schools of Design for the loan of these specimens.' Leeds School of Design Committee Meeting, 27 January 1847, Minutes of the Sub Committee of the Leeds School of Design (1844–54), West Yorkshire Archive Service, Leeds, WYL/23.

[17] This second display appears to have opened between 12 and 4 o'clock and 6 and 9 o'clock on this date alone.

[18] Leeds Mechanics' Institution and Literary Society Committee Meeting, 2 February 1847, Leeds Mechanics' Institution and Literary Society Minute Book (1846–47), p. 208, West Yorkshire Archive Service, Leeds, WYL368/2. See also 'Leeds Government School of Design, in connexion with the Mechanics' Institution', *Leeds Mercury*, 30 January 1847, p. 1.

[19] 'Leeds School of Design', *Leeds Mercury*, 6 February 1847, p. 5.

[20] Council of the School of Design, *Fifth Report of the Council of the School of Design, for the year 1845–46* (London: printed by William Clowes and Sons for Her Majesty's Stationery Office, 1846), p. 16.

scheme was, however, conspicuously lacking in specific details. How far the intentions of the collection were realised remains difficult to discern and is likely to have varied considerably from one location to another. The objects were certainly received with enthusiasm by those with vested interests in the success of the Leeds School of Design. The concept of a circulating collection was formalised in the 1850s, and we will return to this in the final section of this chapter to further consider the implications of these itinerant objects.

Leeds Academy of Arts, 1853–55

While the Leeds School of Design benefited from its umbilical connection to its parent institution in London through access to grants and loans of objects for teaching and public display, it rejected elements of the centralised organisation and curriculum that did not suit its interests and priorities. For example, the committee of the Leeds Mechanics' Institution, the body responsible for the establishment and maintenance of the Leeds School of Design, chose to accept the rules issued by Charles Heath Wilson at the Head School of Design at Somerset House, 'with the exception of Rule 4'.[21] This referred to what was the most divisive regulation: 'No Student to be admitted who is studying Fine Art solely for the purpose of being a Painter or Sculptor.'[22] In order to circumvent this rule without the loss of the crucial annual government grant, many schools listed their students as 'occupation undetermined'.[23] This category was used when the student had no intention of pursuing a career in industrial design or the applied arts.[24] Despite this level of subterfuge there were occasions when it was more prudent for regional schools such as Leeds to emphasise their compliance with the rules that governed their activities. The *Leeds Mercury* quoted the first art master of the new

21 Minutes of Sub Committee of the Leeds School of Design, 1844–54, 23 November 1846.

22 Charles Heath Wilson, *Government School of Design Rules Approved and Adopted by the Council, for Attendance, Conduct, and Studies* (London: printed by W. Clowes and Sons, 1843).

23 Macdonald, *The History and Philosophy of Art Education*, pp. 73–74.

24 For example, 44 of the 180 students registered at the Sheffield School of Design in 1847 were listed as 'occupation undetermined'. This was twice the next largest group of students employed in the chasing and etching trades. See *The Government School of Design, Victoria-Street, Glossop-Road, Sheffield. A Report of the Proceedings at the Annual Meeting* (Sheffield: printed by George Ridge, 1847), p. 4.

Leeds School of Design, Claude Lorraine Nursey (1816–73), on the mode of instruction he intended to employ when teaching commenced:

> My system of education will be the same as that adopted at the head-school at Somerset-house. I shall endeavour to place before my pupils the best and most approved examples from sculptures, consisting of those works that the united judgement of successive ages have handed down to us. First and foremost, the remains of Greek and Roman art must be studied, until a thorough appreciation of their beauties has been imbibed.[25]

Only two years after Nursey had made this statement of conformity, circumstances had already changed. More affluent students, including women, had enrolled at the Schools of Design as an inexpensive alternative to the employment of a private drawing master. Charged a higher fee for their tuition, the middle classes subsidised the system, although their presence undermined the prescribed remit of the schools to produce designers for industry.[26] To court students of a higher social position with artistic aspirations, the Leeds School of Design deliberately situated its curriculum in relation to the more prestigious Royal Academy of Arts:

> Mr. Nursey only follows up the plan adopted at the Royal Academy of Arts, London, (of which Mr. Nursey is a life student), where they compel all students entering as sculptors, to keep a considerable time in the Antique school, drawing outline in chalk before they are allowed to model.[27]

Although his allegiance appeared to have shifted, both of Nursey's statements referred to essentially the same mode of education, centred

[25] Claude Lorraine Nursey, 'Leeds School of Design', *Leeds Mercury*, 12 December 1846, p. 8.

[26] The fees at the Leeds School of Design were lowest for the evening classes intended for the working classes. The Morning Class charged 8s. per quarter, the Afternoon Class for Females 8s. per quarter and the Evening Class 6s. per quarter. Students who were not also members of the Mechanics' Institution were charged an additional 2s. per quarter. Each class was two hours long and ran five days per week. From 14 January 1850, a universal fee of 2s. per month was charged, in line with other branch Schools of Design.

[27] 'Leeds Government School of Design', *Leeds Mercury*, 11 November 1848, p. 1.

on drawing from plaster casts of canonical antique exemplars. Recourse to the rhetoric of the academic tradition was connected to a growing dissatisfaction with the structure, syllabus and results of the Schools of Design. The network of branch schools was generally considered to have failed to supply skilled labour to local industry, as the majority of students left after the elementary stages with only a basic command of drawing historic ornament in outline. Neither did the schools satisfy those with artistic ambitions, as the provision for tuition in life drawing, landscape studies and painting was limited. It was in this climate that an attempt was made in 1852 to establish a branch of the Royal Academy of Arts in Leeds, as a means of establishing a tangible boundary between fine art and the application of art to industry.[28]

The idea was conceived by the local artist Richard Waller (1811–82) and his associates during the winter of 1852, and from its inception it was intended that the Leeds Academy of Arts would include 'a School of Art adapted to a higher purpose that is usually found in Schools of Design'.[29] In a statement to the *Leeds Mercury*, it was declared that:

> The primary object of the Academy is the general improvement of art in the locality. In the plan of its promoters, it is not viewed as a rival institution to the School of Design, but as an aid to that admirable society, in furthering the study of the higher branches of art, to which the School of Design may be considered as preparatory; and by thus encouraging a taste for the higher productions of art, it is intended to open an access by artistic merit to provincial honours, capable of raising to due estimation those whose talents entitle them to distinction.[30]

A number of local societies and institutions had shared this 'primary object' during the first half of the nineteenth century, but the Leeds Academy of Arts was the first to attempt to provide a specific training in the fine arts for graduates of the Leeds School of Design and an infrastructure to support artists in their own region.

28 'Leeds Academy of Art', *Leeds Mercury*, 23 October 1852, p. 7.

29 Leeds Academy of Arts, *First Annual Report of the Committee of the Leeds Academy of Arts, with List of Subscribers* (Leeds: printed by J. Buckton, 1853), p. 4.

30 'Leeds Academy of Art', *Leeds Mercury*, 23 October 1852, p. 7. Waller could hardly have advertised the Leeds Academy of Arts as a rival institution to the Leeds School of Design in the *Leeds Mercury*, as it was edited by Edward Baines Junior, who had been instrumental in the establishment and administration of the Leeds School of Design.

The committee secured the endorsement of the current president of the Royal Academy of Arts in London, Sir Charles Eastlake (1793–1865), who was appointed honorary president of the Leeds Academy of Arts. In addition, the majority of academicians and associates of the Royal Academy agreed to be honorary members of this novel regional outpost.[31] The Leeds Academy was modelled on its parent institution insofar as it was to incorporate a school and hold exhibitions to generate income. To secure additional support, the Academy appropriated the familiar economic logic of the Schools of Design by claiming that a knowledge of form and colour was 'indispensably necessary for excellence in the design and execution of all descriptions of what are called fancy goods ... the basis of our manufacturing superiority, in many important respects, over the most skilful of our foreign competitors'. Through this instrumental relationship between art and industry the Academy attempted to solicit the support of 'the *élite* of this town and neighbourhood, and especially from our Merchant-Princes and Opulent Manufacturers'.[32]

However, the 'Merchant-Princes and Opulent Manufacturers' offered only limited patronage, and as a result, the Leeds Academy of Arts was active for less than three years before it was reconstituted as an equally short-lived art union.[33] The first annual exhibition, held between June and September 1853, was considered a success, if not as profitable as had been hoped.[34] Premises at 8 Bond Street, described as 'spacious buildings in the most central part of town', were rented and furnished to accommodate both the school and its exhibitions.[35] The debt incurred by the Academy during this process was not cleared by its first exhibition.[36] A second exhibition was hastily arranged as a solution to its precarious financial position. The attendance, however, was not as high as the first exhibition, and Waller admitted that 'the winter exhibition was got up at a most serious loss to the Academy'.[37] The

31 Leeds Academy of Arts, *First Annual Report*, p. 6.

32 Leeds Academy of Arts, *First Annual Report*, pp. 5–6.

33 'The Yorkshire Art Union', *Leeds Mercury*, 26 May 1855, p. 8.

34 'The local artists and lovers of art in Leeds, formed at this time a public exhibition of modern works of art, styled "The Leeds Academy of Arts." The aggregate value of the paintings exhibited was said to be upwards of £20,000. The exhibition was open three months.' John Mayhall, *The Annals and History of Leeds, and other places in the county of York, from the Earliest Period to the Present Time* (Leeds: Joseph Johnson, 1860), p. 631.

35 'Leeds Academy of Arts.—Art Union', *Leeds Mercury*, 15 January 1853, p. 8.

36 Leeds Academy of Arts, *First Annual Report*, p. 7.

37 Richard Waller, 'Leeds Academy of Arts', *Supplement to the Leeds Mercury*, 28 October 1854, p. 12.

third and final exhibition, held during the autumn of 1854, resulted in a scandal from which the Academy never recovered. After the exhibition had closed, the pictures due to be returned to London were impounded by the Great Northern Railway Company because the bill of £49 for the transport of the works displayed at the previous exhibition had not been paid.[38] The *Athenæum*, a weekly literary journal published in London, picked up the story in April 1855:

> The Artists, at least, as a body, will derive a salutary warning from the fact, and learn, when one or more unknown individuals introduce themselves into their studios, with specious demeanour, with a printed form containing a goodly array of names of persons, not a fraction of whom are interested in the subject of fine art, with a plausible prospect of a local Art Union—in the hope of cultivating a taste for the arts in their district, already in possession of that universal *panacea* for the production of provincial artizan progress—a School of Design—the artist of London will, it is to be hoped, be on his guard ere he rashly confide his property to plausible and specious persons who early in their interviews address themselves to his pecuniary interests.[39]

The double use of the term 'specious' suggests that the artists had been deliberately misled by the unnamed representative of the Leeds Academy of Arts, although it is difficult to determine whether they acted in bad faith or through blind optimism. That the journalist also referred to them as 'unknown individuals' demonstrates that they operated outside metropolitan art networks.

The 140 artists whose work had been seized arranged to meet in order to resolve the dispute. Legal action was suggested, but it was decided that 'law, few artists have a taste for' and it would be cheaper to pay the amount that was owed between them.[40] This action proved sufficiently embarrassing for a subscription to be instigated in Leeds to clear the debt and 'to prevent a public disgrace and severe individual injury from being entailed by that Exhibition'.[41] Sixteen cases of pictures were released back to their owners in May 1855, almost a year after they had been sent for exhibition, which marked the dissolution of the Leeds

38 'The Borrowed Pictures', *Bradford Observer*, 3 May 1855, p. 5.

39 'Fine Art Gossip', *Athenæum*, 28 April 1855, pp. 494–95.

40 R.W. Buss, 'The London Artists & The Leeds Academy', *Supplement to the Leeds Mercury*, 19 May 1855, p. 1.

41 'Fine Arts' Exhibitions', *Supplement to the Leeds Mercury*, 5 May 1855, p. 1.

Academy of Arts.[42] The *Leeds Mercury* lamented: 'this department of art must relapse into the sleep from which the abortive attempt of the Academy has failed to arouse it'.[43]

Although the Leeds Academy of Arts benefited from the nominal endorsement of the Royal Academy of Arts in London, a Mill Hill location close to the cultural and intellectual centre of the town, and a legacy of local support for the visual arts over the preceding decades, it did not secure enough patronage from the influential middle-class elite. It is possible that those associated with the campaign to establish a School of Design in Leeds viewed the Academy as unnecessary at best and a direct rival at worst, although there is no evidence that the Academy enrolled or taught any students during its short life. As an artist-led organisation, it could not compete with the local and national organisational structures that supported the Leeds School of Design, in the form of the Leeds Mechanics' Institution and Head School of Design in London.

Unlike the artists who founded the Royal Academy of Arts almost a century earlier, the artists associated with the Leeds Academy of Arts had limited reputations and were not generally of sufficiently high social status to look to their peers for patronage. Neither did Leeds have a tradition of artist-led activity; as Trevor Fawcett noted in his study of art outside London, even the Northern Society for the Encouragement of the Fine Arts had been established by 'gentry and amateurs' rather than professional artists.[44] The private schools of art that emerged in the second half of the eighteenth century—many of which were intended to prepare students for successful entry into the Royal Academy Schools—were concentrated in London and already past the height of their popularity when the Leeds Academy of Arts was founded.[45] Beyond London, the societies and institutes that provided academic tuition in the fine arts began to close or reconstitute their provision as Government Schools of Design. The Royal Academy of Arts never operated as an official franchise with regional branches, so the Leeds Academy had no recourse to its metropolitan namesake to bolster its precarious position. The humiliating collapse of the Academy undermined the cultural confidence of the town and lingered in its

42 'Leeds Academy of Arts. The Detained Pictures', *Leeds Mercury*, 26 May 1855, p. 8.

43 'Fine Arts' Exhibitions', p. 1.

44 Trevor Fawcett, *The Rise of English Provincial Art: Artists, Patrons, and Institutions outside London, 1800–1830* (Oxford: Clarendon Press, 1974), p. 168.

45 Macdonald, *The History and Philosophy of Art Education*, p. 33.

memory. It would be over a decade before Leeds attempted to stage another exhibition dedicated to the fine arts, and in the meantime, it was left to the Leeds School of Design to continue to hold temporary public exhibitions of the applied arts.

The 'Travelling Museum' at the Leeds School of Practical Art, 1855–56

For forty-one days between 24 November 1855 and 5 January 1856, the Leeds School of Practical Art—as the Leeds School of Design had been renamed in 1853—exhibited the Circulating Collection, or 'Travelling Museum' of the Museum of Ornamental Art at Marlborough House in London.[46] Examples of Italian majolica, German and Flemish stoneware, Chinese porcelain, Russian engraving, French furniture, Turkish and Indian silks and Spanish and Irish lace embodied both the anxiety and possibility represented by foreign manufactures for British designers.[47] The collection was part of an attempt to define and diffuse an international, standardised statement of the principles of correct design, informed by the practice and pedagogy of educational institutions in Paris, Lyons, Berlin and Munich. The regulations for the 'Travelling Museum' stipulated that the regional recipient must also assemble its own local loan collection to supplement the exhibition.[48] This confluence of objects provides an opportunity to compare regional, national and international priorities in the contexts of mid-nineteenth-century British art and design education.

The Circulating Collection was devised by the Department of Science and Art in 1854 and operational by February 1855 as 'a faithful abstract or abridgement' of the Marlborough House Museum in London.[49] The stated

46 'Leeds School of Practical Art. Great Exhibition & Conversazione Last Night', *Leeds Mercury*, 20 November 1855, p. 3.

47 J.C. Robinson, *Catalogue of a Collection of Works of Decorative Art: Being a Selection from the Museum at Marlborough House, Circulated for Exhibition in Provincial Schools of Art*, 5th edn (London: printed by George E. Eyre and William Spottiswoode for Her Majesty's Stationery Office, 1856), pp. 3–6.

48 Circular from the Board of Trade dated 11 August 1854, condition number two: 'That the committee of the school endeavour to add to the exhibition by obtaining loans of specimens from the collections of private individuals in the neighbourhood.'

49 Department of Science and Art, *Third Report of the Department of Science and Art* (London: printed by George E. Eyre and William Spottiswoode for Her Majesty's Stationery Office, 1856), p. 70.

aim of the collection was 'the illustration, by actual monuments, of all art which finds its material expression in objects of utility, or in works avowedly decorative'.[50] The resulting 'Travelling Museum' also operated as a way to appease the regional bodies that had contributed to the Great Exhibition of 1851 but had not directly benefited from its substantial profits.[51] In a letter to the Royal Commissioners for the Great Exhibition, James Kitson, in his capacity as president of the Leeds Mechanics' Institution and Literary Society, expressed the opinion that surplus funds should be directed towards the support of regional mechanics' institutes and schools of design. Kitson argued that these bodies provided 'a sound, intellectual culture' and a means of diffusing taste through art training, towards the improvement of national industry.[52] The Department of Science and Art, however, was too invested in centralised, metropolitan authority to distribute the profits of the Great Exhibition to the regions directly, even though their objectives were broadly consistent. Part of the role of the Circulating Collection was to present the Department of Science and Art as a progressive and generous body, which claimed:

> for the first time perhaps in the history of museums of rendering moveable the treasures acquired, and of bringing home to the millions of the land opportunities for the study of the beautiful in art, which have hitherto, at least in some degree, been the privilege only of dwellers in metropolitan cities. This intention it is thought may be carried out with little injury to the Museum as a unity, and with great gain in the direction of the chief object of its foundation.[53]

The collection was first displayed in Birmingham between 26 February and 12 April 1855, and on 3 June 1855 the secretary of the Leeds School of Practical Art was instructed to write to both the Birmingham School and Marlborough House to determine the expenses likely to be incurred and how much space the collection required for public display.[54] By

50 Robinson, *Catalogue of a Collection of Works of Decorative Art*, p. 4.

51 Department of Science and Art, *Third Report of the Department of Science and Art*, p. 70. The Leeds School of Practical Art was the fifth school to receive the collection, after Birmingham, Nottingham, Macclesfield and Norwich.

52 A loose letter dated 20 October 1851 from the Leeds Mechanics' Institution and Literary Society Minute Book (1851–55), West Yorkshire Archive Service, Leeds, WYL368/4.

53 Robinson, *Catalogue of a Collection of Works of Decorative Art*, p. 4.

54 Leeds School of Practical Art Committee Meeting, 3 June 1855, School of Art Minute Book (1854–68), West Yorkshire Archive Service, Leeds, WYL368/26.

November 1855 these questions had been answered: the cost of staging the exhibition was estimated at £49 2s. 9d. and a warehouse adjoining the School had been secured without charge for one month through William Beckett.[55] Hosting the 'Travelling Museum' was perceived by the committee of the Leeds Mechanics' Institution and Literary Society as an opportunity to improve its financial position.[56] Economic considerations were especially important to the Leeds School of Practical Art under the reforms imposed by the Department of Science and Art. The Leeds School had been selected in 1853 to take part in an experimental financial arrangement, because it had spent the most on the instruction of each student in 1851.[57] To encourage the School to subsidise its own activities, £50 of the art master's salary of £200 was to come directly from student fees, where it had previously been paid in full through the system of central funding.[58] After a year under the new structure it was reported that 'a gratifying result has attended this arrangement. The school was inspected early in December [1854], and found to be

55 Special Meeting of the Committee, 8 November 1855, Leeds Mechanics' Institution and Literary Society Minute Book (1855–59), West Yorkshire Archive Service, Leeds, WYL368/5, p. 90, and Leeds School of Practical Art Committee Meeting, 3 November 1855, School of Art Minute Book (1854–68), West Yorkshire Archive Service, Leeds, WYL368/26. William Beckett (1784–1863) was a partner in the banking firm Beckett & Co. and a Conservative MP for Leeds (1841–52) and Ripon (1852–57). Beckett supported the voluntary societies of Leeds, including contributions towards the Philosophical Hall, the Leeds Institute building and the new Infirmary. As a member of the committee of the Mechanics' Institution, Beckett helped to secure a grant to establish the Leeds School of Design, having gained Parliamentary support from William Gladstone and practical advice from George Birkbeck, founder of the London Mechanics' Institute. Beckett frequently chaired the *conversazione* of the School and donated objects to its exhibitions, including specimens of French manufactures. See also Richard Vickerman Taylor, *The Biographia Leodiensis; or, Biographical Sketches of the Worthies of Leeds and Neighbourhood, from the Norman Conquest to the Present Time* (London: Simpkin, Marshall & Co.; Leeds: John Hamer, 1865), pp. 506–9.

56 Special Meeting of the Committee, 8 November 1855, Leeds Mechanics' Institution and Literary Society Minute Book (1855–59), West Yorkshire Archive Service, Leeds, WYL368/5, p. 88.

57 Department of Science and Art, *First Report of the Department of Science and Art* (London: printed by George E. Eyre and William Spottiswoode for Her Majesty's Stationery Office, 1854), p. xxxiv. The Department calculated that the Leeds School had spent £10 11s. 2d. on the instruction of each pupil.

58 Department of Science and Art, *First Report of the Department of Science and Art*, p. xxxv.

in a most efficient state. Leeds is, therefore, no longer a grant school, supported by State assistance.'[59]

Despite its presentation as a good news story by the body that benefited from the cost saving, the withdrawal of the grant exacerbated the debts accrued by the School, and so the decision to stage a public exhibition was at least partially pragmatic.[60] The Department of Science and Art determined that students should be admitted to the exhibition free of charge, 'but that all other persons, not students, pay a moderate fee for admission, which should be higher in the morning than the evening'.[61] It was further stipulated that the profits should be used towards the expenses of the exhibition, including the transport of the objects, before 'the balance be appropriated in the following proportions, namely; one quarter to the masters fee-fund; one-half to the purchase of examples for a permanent museum, &c; and one quarter to the general fund of the school'.[62] These regulations were waived for the Leeds School, however, as during a meeting between Thomas Wilson as chairman of the Leeds School of Practical Art and Henry Cole, it was recorded that 'Mr. Cole assented to the request of the Committee that the proceeds of the Exhibition ... be employed to liquidate the debt of the School'.[63] Notwithstanding the potential

[59] Department of Science and Art, *Second Report of the Department of Science and Art* (London: printed by George E. Eyre and William Spottiswoode for Her Majesty's Stationery Office, 1855), p. xx. John Charles Swallow arrived in Leeds on 1 September 1854 and was officially appointed headmaster of the Leeds School of Practical Art on 4 September 1854 to replace Thomas Gaunt.

[60] From a letter by Thomas Wilson as chairman of the Leeds School of Practical Art to solicit further subscriptions: 'From its opening, until a very recent period, the School has enjoyed a Government Grant of not less than £150. That Grant has been withdrawn for more than a year, and there is no prospect of its renewal; the Central Department having decided to leave the Provincial Schools to be supported by *local aid*. The Committee have now carried the School on for more than a year with the aid of a very small Subscription, but the result having been the accumulation of a debt of more than £100.' Committee Meeting, 25 November 1855, Leeds School of Art Committee Minute Book (1854–68), West Yorkshire Archive Service, Leeds, WYL368/26.

[61] Copy of a Circular to the Local Schools of Art on the Regulations for the Exhibition of the Circulating Collection, 11 August 1854, in Robinson, *Catalogue of a Collection of Works of Decorative Art*, p. 2.

[62] Copy of a Circular to the Local Schools of Art on the Regulations for the Exhibition of the Circulating Collection, p. 2.

[63] Leeds School of Practical Art Committee Meeting, 15 December 1855, School

to return a profit, the expenses associated with the the collection were significant, as were the financial risks. The host institution was responsible for the cost of transporting the objects from London, in 'a carriage or truck, constructed especially for the purpose, and adapted to travel on all railways'.[64] The possibilities created by this mode of transport had been noted by the council of the School of Design in its annual report of 1845, which recorded that all the current branch schools were connected by rail.[65]

The Circulating Collection also had to be insured and protected for the duration of its display.[66] There was a palpable concern for the security of the objects, most particularly for the Sèvres porcelain and fragile examples of china loaned by Queen Victoria (Figure 4.1). These examples stimulated intense local interest in Leeds, particularly relative to their value, with one vase believed to be worth £1,000 and another £4,000.[67] The *Leeds Mercury* were particularly effusive about the perceived benefits of this indirect royal patronage:

> Beyond the intrinsic value and great beauty of this portion of the exhibition, the mere fact that the Queen had permitted the Department of Science and Art to select from her private collection whatever they deemed most useful for their purpose, and circulate them for exhibition amongst the various provincial Schools of Art, was of itself sufficient to ensure for the whole a large share of interest and attention. And could her Majesty have heard the warm and hearty praise which this gracious act elicited from the

of Art Minute Book (1854–68), West Yorkshire Archive Service, Leeds, WYL368/26.

64 'General Arrangements for Exhibition in Provincial Schools of Art', in Robinson, *Catalogue of a Collection of Works of Decorative Art*, p. 7.

65 Council of the School of Design, *Fourth Report of the Council of the School of Design, for the year 1844–45* (London: printed by William Clowes and Sons for Her Majesty's Stationery Office, 1845), p. 21.

66 The committee of the Leeds School of Practical Art insured the collection to the value of £2,000 with the Leeds and Yorkshire Company. Two men were employed to guard the exhibition during the day and evening alongside a policeman, with a separate watchman for each Sunday. Leeds School of Practical Art Committee Meeting, 21 and 25 November 1855, School of Art Minute Book (1854–68), West Yorkshire Archive Service, Leeds, WYL368/26.

67 These valuations appear to have first been brought to public attention when the royal collection of Sèvres porcelain was displayed as a temporary loan exhibition at Marlborough House in 1852. See, for example, 'Sèvres and other Porcelains at the Exhibition of Art Manufactures', *Illustrated London News*, 18 September 1852, p. 221.

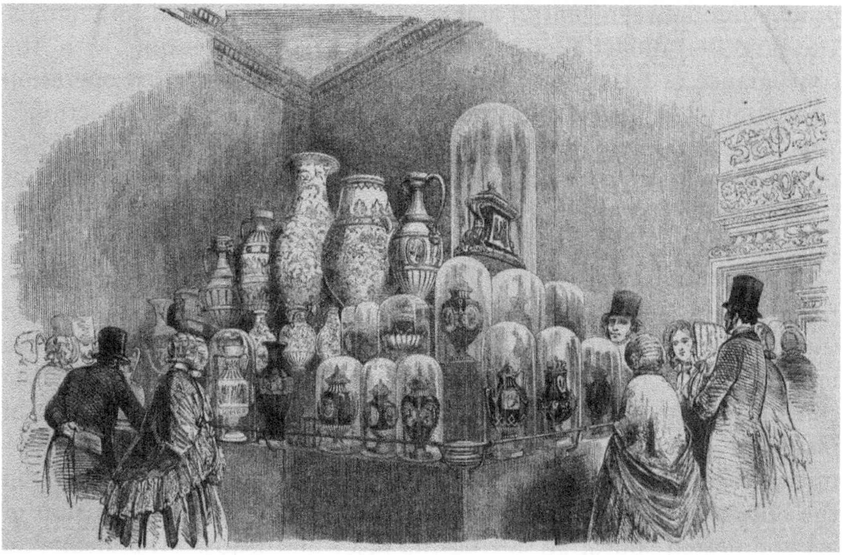

Figure 4.1 Anonymous, *Sèvres China from the Museum of Art Manufactures, at Marlborough House*, from *The Illustrated London News*, 18 September 1852, wood engraving. Collection of the author

> throng of visiters [*sic*] during the evening, we are sure she would have felt amply repaid for her condescension.[68]

In addition to employing police officers and watchmen to guard the exhibits, Charles B. Worsnop of the Department of Science and Art accompanied the Circulating Collection to each location.[69] Worsnop also delivered daily lectures on the collection, which were supplemented by lectures given 'twice a week by a Leeds gentleman in the fine arts section'.[70] This arrangement illustrated the negotiation of knowledge

[68] 'Leeds School of Practical Art. Great Exhibition & Conversazione Last Night', *Leeds Mercury*, 20 November 1855, p. 3.

[69] Worsnop was a clerk for the Department of Science and Art, who was later appointed Assistant Keeper of Museum Collections at the South Kensington Museum. The regulations stipulated that: 'An officer of the Department accompanies the collection and remains in charge during the period of its each locality.' J.C. Robinson, 'General Arrangements for Exhibition in Provincial Schools of Art', in *Catalogue of a Collection of Works of Decorative Art*, p. 7.

[70] 'The Exhibition of the Leeds School of Practical Art', *Leeds Mercury*, 24 November 1855, p. 5.

production between central and regional institutions and individuals. The proposed model appeared to offer a balanced solution, with 'the continuance or extension of the system of Government co-operation now attempted depending entirely on correspondent local action'.[71] This 'local action' was intended to stimulate the formation of regional museums with composite collections of copies from the Museum of Ornamental Art at Marlborough House, augmented with objects of local interest and importance.[72]

The purchase of photographs, electrotypes and gelatine casts taken from objects held by the museum in London was an extension of the supply of plaster casts and other forms of reproductions to the branch Schools of Art under the Department of Science and Art. The concurrent encouragement to form regionally specific collections of historical and contemporary industrial manufactures appeared to mitigate the centralised, standardised direction that had characterised the relationship between London and the regions in the provision of art and design education during the 1830s and 1840s.[73] The display of the Circulating Collection in Leeds did not, however, directly lead to the formation of a permanent museum of the sort envisaged by the Department of Science and Art.[74] The extent to which the exhibition of the Circulating Collection was mediated by local interests and particularities is open to question, as the department had established stringent procedures to ensure uniform presentation and interpretation wherever it was staged. The spatial arrangement of the objects was predetermined by five large glazed cases and seventy glazed frames, which protected

71 Robinson, *Catalogue of a Collection of Works of Decorative Art*, p. 6.

72 Robinson, *Catalogue of a Collection of Works of Decorative Art*, p. 5.

73 'But in reference to local endeavours, an undertaking of the greatest importance in a practical point of view would be the formation of historic collections of any one of our great industries, on the very spot of its development, where alone the requisite materials and illustrative knowledge could be gleaned... Nor should art industry of the present period be forgotten; in every town the "chefs d'œuvre" of the day, accompanied with a due record of designers, manufacturers, and even skilled workmen associated in their production, should meet with a public and abiding recognition in the place of their manufacture.' Robinson, *Catalogue of a Collection of Works of Decorative Art*, p. 5.

74 Arguably, this was because Leeds already had an established culture of temporary polytechnic and fine art exhibitions associated with the Mechanics' Institution and the Northern Society for the Encouragement of the Fine Arts respectively. The museum of the Leeds Philosophical Society also fulfilled part of this remit from 1821 until the collection was transferred to Leeds City Council in 1921.

the collection during transport and also provided the means of display.[75] In addition to the constant presence of an officer from the Department of Science and Art, whose secondary function was to provide public lectures, the interpretation of the objects was also controlled by the department through the labels and catalogue which accompanied the collection, alongside the classifications imposed on it. This dominance of knowledge persisted through the archive; the official catalogue remains the most substantial account of the collection and necessarily informs subsequent investigations into its content and meaning.

The *Leeds Mercury* recorded the local donors to the exhibition with a limited description of the objects loaned, which included 'a very imposing array of paintings, prints, sculpture, bronzes, gold and silver articles, and specimens of decorative art in paper, carpeting, &c., contributed by tradesmen and private gentlemen resident in Leeds'.[76] However, the details of these donations were not recorded, nor discussed at any length. The only comparison between the different sections of the exhibition was with the rooms of the School of Practical Art, which were described as 'filled with sculpture' but 'less attractive than any other portion of the Exhibition'.[77]

Conclusion

After the 'Travelling Museum' had moved on to Sheffield, the Leeds Mechanics' Institution and Literary Society issued a celebratory account of the exhibition as part of its annual report:

> The really successful Art Exhibition which was lately instituted in connection with the School, and which your Committee hope will be productive of great and varied good, is another and safe illustration, within a short space, of the necessity of carrying a really enterprising spirit into matters of education. When the scheme of an Exhibition was first mooted, and took the form of hazardous and expensive estimates, many friends of the School were doubtful of its success and utility, but by perseverance and economy on our part, met by the public spirit and liberality

75 Robinson, *Catalogue of a Collection of Works of Decorative Art*, p. 7.

76 'Leeds School of Practical Art. Great Exhibition & Conversazione Last Night', *Leeds Mercury*, 20 November 1855, p. 3.

77 'Leeds School of Practical Art. Great Exhibition & Conversazione Last Night', p. 3.

> of many of our townsmen, we have succeeded in meeting all claims and providing for our people a cheap, valuable, and highly instructive display.[78]

Despite this highly favourable appraisal and large number of visitors, the exhibition failed to return the profit that had been so sorely required for the Leeds School of Practical Art. Although the total receipts exceeded those of all four of the previous locations in which the collection had been displayed, it was reported that the profit of £115 was almost matched by the expenses.[79] Despite the poor margin, the Circulating Collection conferred cultural capital on Leeds and its School of Practical Art, in the same way that the first teaching collection associated with the School of Design had been so crucial as a marker of the intellectual ambitions of the town. Diminished central funding from the Department of Science and Art had made the Leeds School of Practical Art—and by extension its parent organisation, the Leeds Mechanics' Institution and Literary Society—more reliant on generating their own income through student fees and exhibition tickets. The influence of South Kensington remained strong during this period and the combination of loaned objects from London displayed alongside local loans would continue to inform the practice of exhibition making in Leeds for the rest of the century.

Following a strong start in 1839, the exhibitions held in Leeds over the course of the next fifteen years tended to generate meagre profits or, in the case of the Leeds Academy of Arts, embarrassing losses. Disappointing financial results were not the only metric used to assess the significance of these events; the support of audiences and the educational, social and cultural value of the exhibitions were generally emphasised regardless of the balance sheet. In addition to the

[78] Leeds Mechanics' Institution and Literary Society, *Annual Report of the Committee of the Leeds Mechanics' Institution and Literary Society* (Leeds: printed by John Kershaw and Son, 1856), pp. 14–15.

[79] 'Close of the Leeds School of Practical Art Exhibition', *Leeds Mercury*, 8 January 1856, p. 4. The disappointing returns did not prevent the Leeds School of Practical Art from hosting the Circulating Collection for a second time between December 1862 and January 1863. Only Dublin, Liverpool, Leeds and Exeter hosted the Circulating Collection twice. See 'Summary of the General Results of the Circulating Museum from its institution in February 1855 to June 1863 inclusive', Appendix H to the *Eleventh Report of the Science and Art Department of the Committee of Council on Education* (London: printed by George E. Eyre and William Spottiswoode for Her Majesty's Stationery Office, 1864), p. 150.

factors specific to the collapse of the Leeds Academy of Arts, exhibition organisers consistently expressed disproportionate confidence in the prospects of a given event, coupled with a tendency to erode profits through high expenditure. It is likely that civic rivalry was at least a partial cause of this immoderate approach, exacerbated by the inter-city competition encouraged by the Department of Science and Art's publication of visitor figures and profits from exhibitions across the network of regional schools. The prevailing belief that Leeds should be able to exceed the achievements of its regional peers frequently led to finances being overextended, but equally resulted in an ambitious sensibility that would not accept provincial status.

5

The National Exhibition of Works of Art at Leeds, 1868

The National Exhibition of Works of Art was one of the largest and most significant temporary public exhibitions to be staged in Britain in the second half of the nineteenth century. Held between 19 May and 31 October 1868, 570,000 people attended an ambitious display of more than 6,000 objects, contributed by over 800 individuals and institutions.[1] The exhibition was intended to fulfil a number of practical functions: to raise funds to complete a new General Infirmary building, to elevate Leeds's cultural status and to instigate a permanent art gallery for the town. These anticipated outcomes had motivated the staging of public exhibitions in Leeds before, but the extent of the National Exhibition of Works of Art was unprecedented in Yorkshire and had not been seen in Britain since the Art Treasures Exhibition in Manchester in 1857, to which it was most often compared.[2] Despite its scale and significance, the exhibition has not yet been the subject of substantial scholarship, unlike many of its regional and national equivalents. This omission in the literature on exhibition histories is surprising, but may have been at least in part caused by the failure of the exhibition to generate a profit and a permanent legacy in the form of a public art gallery. These narrow measures of success belie the wider importance of the exhibition. It connected Leeds to new and established networks of collection and display, provided the town with a national platform and advanced arguments in favour of the formation of a permanent art gallery. It may have taken a further twenty years to be realised, but as we will witness in Chapter 7, the exhibition of 1868 was crucial to the eventual establishment of Leeds Art Gallery in 1888.

1 Figures quoted in 'Country News', *Illustrated London News*, 7 November 1868, p. 439, and 'Close of the Leeds Exhibition', *The Times*, 2 November 1868, p. 5.

2 'The Leeds New Infirmary. Proposed Fine Art and Industrial Exhibition', *Leeds Mercury*, 28 September 1866, p. 4.

Leeds General Infirmary

Longstanding rivals across the dividing hills of the Pennines, the National Exhibition of Works of Art was intended to elevate Leeds's cultural status to match, if not exceed, that of Manchester. Whereas the Art Treasures Exhibition was accommodated in purpose-built, temporary glass and iron halls outside the centre of Manchester, in Leeds the architectural context was predicated on the practical purpose of the exhibition to support the completion of a new Infirmary building in the middle of the town. The existing Leeds Infirmary had been designed in the Palladian style by John Carr (1723–1807) in 1767. A century later even the extended building had become too small for a population that had grown from 53,612 in 1801 to 217,149 by 1861.[3] The Leeds Mechanics' Institution and Literary Society had also outgrown its rented accommodation, and its new building, designed by Cuthbert Brodrick, had been under construction since 1865. These two institutions were united by a common shortfall in their respective building funds, alongside two building committees with a shared membership and set of interests and priorities. The building committee of the Infirmary had budgeted £30,000 for the new Infirmary building, but when over twice that amount was raised in subscriptions, the committee in turn raised their aspirations and engaged one of the most renowned architects of the Gothic Revival, George Gilbert Scott (1811–78), to design an Infirmary on the most modern principles. The 'pavilion' plan advocated the physical separation of wards from one another to increase light and the circulation of fresh air, thought to reduce the transmission of disease and promote recovery.[4] The foundation stone was laid on 29 March 1864 by the chairman of the building committee, the second-generation industrialist and politician James Kitson (1835–1911), who would go on to join the executive committee of the exhibition.[5]

Despite reducing the degree of ornamentation in Gilbert Scott's design, the cost of construction quickly exceeded even the increased budget and left a deficit of approximately £25,000. In September 1866 the building committee decided to hold an exhibition of fine art

[3] Figures quoted in 'Leeds', *Graphic*, 18 July 1885, pp. 73–79.

[4] 'The New Infirmary at Leeds, Now Occupied by the National Art-Exhibition, Opened by the Prince of Wales on Tuesday', *Illustrated London News*, 23 May 1868, p. 512.

[5] *The General Infirmary at Leeds: Its Work and Progress, 1767–1911* (Leeds: n.p., 1912?), p. 12.

inside the new Infirmary once it had been completed, but before it was furnished as a functional hospital. Although it was likely that the building would be finished the following year in 1867, the committee decided to delay the event until 1868 to avoid clashing with the second Exposition Universelle in Paris.[6] The Chief Commissioner of the exhibition, John Burley Waring (1823–75), planned to visit Paris with a view to obtaining objects for Leeds, although he was cautious in his approach and reported: 'I do not think it advisable to expect much or to draw largely from that source.'[7] Waring had trained as an architect and had written a series of architectural guidebooks to the Crystal Palace at Sydenham with Matthew Digby Wyatt (1820–77). He had been the superintendent of sculpture and ornamental art at the Manchester Art Treasures Exhibition and of the architectural gallery at the 1862 International Exhibition at South Kensington.

In order to galvanise support for the exhibition, Waring wrote and published a detailed report in March 1867. His argument was highly effective in securing offers of objects, expertise and contributions to the guarantee fund, although some of his claims were contested. Waring stated that Great Britain was entirely without the public galleries of art so common in Europe, an omission he described as 'a real and serious subject of reproach to this country'.[8] The *Glasgow Herald* took exception to this assertion and criticised Waring's ignorance of the following facts:

> that the city of Glasgow has been the first provincial city in the United Kingdom to establish a public gallery of art; that Glasgow has done this entirely without Government aid of any kind ... that Glasgow is still forming by her own unaided resources galleries for art purposes only equalled in the Metropolis.[9]

The article referred to the McLellan Galleries acquired by Glasgow Corporation after the death in 1854 of Archibald McLellan, who had bequeathed his significant collection of old master paintings to the city. The *Glasgow Herald* concluded the article in more conciliatory tone:

[6] 'The Leeds New Infirmary. Proposed Fine Art and Industrial Exhibition', *Leeds Mercury*, 28 September 1866, p. 4.

[7] J.B. Waring, *Extract from the First Report by J.B. Waring, Esq., General Manager and Chief Commissioner of the "National Exhibition of Works of Art," to be held in the New Infirmary at Leeds, in 1868* (Leeds: the Committee, 1867), p. 6.

[8] Waring, *Extract from the First Report by J.B. Waring*, p. 2.

[9] 'Galleries of Fine Art and Exhibition at Leeds, in the new Infirmary Building', *Glasgow Herald*, 11 May 1867, p. 7.

> We trust that our Leeds friends, to whose project we wish every success, will pardon us for pointing out that they have in their statement overlooked the just claims of this city to have taken the first step in the direction which they are following. We venture to say that they may profit both by our errors and by our example.[10]

The forceful and public correction of Waring's misapprehension demonstrated what was at stake for towns and cities engaged in commercial and cultural competition. Glasgow Corporation had invested the huge sum of £44,500 to purchase the McLellan Galleries and was understandably keen to have its position as the first city to open a civic art gallery recognised.[11]

Looking to London

Waring's exclusion of Glasgow from his discussion points towards a metropolitan bias. He was keen to guard against accusations of provincial philistinism and formed a 'London Committee of Advice', including representatives from the National Portrait Gallery, Society of Antiquaries, Royal Academy of Arts, Royal Institute of British Architects, Sir John Soane's Museum, Crystal Palace, British Museum, South Kensington Museum and the National Gallery, to 'assure to the Exhibition a national character, [and] to guarantee the owners of works of art that the very greatest experience, knowledge, and good taste, as well as the most practical counsel, shall be brought to bear upon the collection, arrangement, and preservation of those treasures of painting and sculpture on which their owners justly set so great a value' (Figure 5.1).[12] The recruitment of metropolitan museum professionals, collectors and art historians did not, however, diminish the role of local individuals and voluntary societies, nor did it erase their priorities. The entire executive committee was composed of second- and third-generation industrialists whose economic, political and cultural activities were embedded in the town. The experience of the Leeds Mechanics'

[10] 'Galleries of Fine Art and Exhibition at Leeds, in the new Infirmary Building', p. 7.

[11] G. Smith, 'McLellan, Archibald (1795/1797–1854), art collector and local politician', *Oxford Dictionary of National Biography* (Oxford University Press, 2004), https://www-oxforddnb-com.libproxy.york.ac.uk/view/10.1093/ref:odnb/9780198614128.001.0001/odnb-9780198614128-e-17663 (accessed 18 February 2022).

[12] Waring, *Extract from the First Report by J.B. Waring*, p. 3.

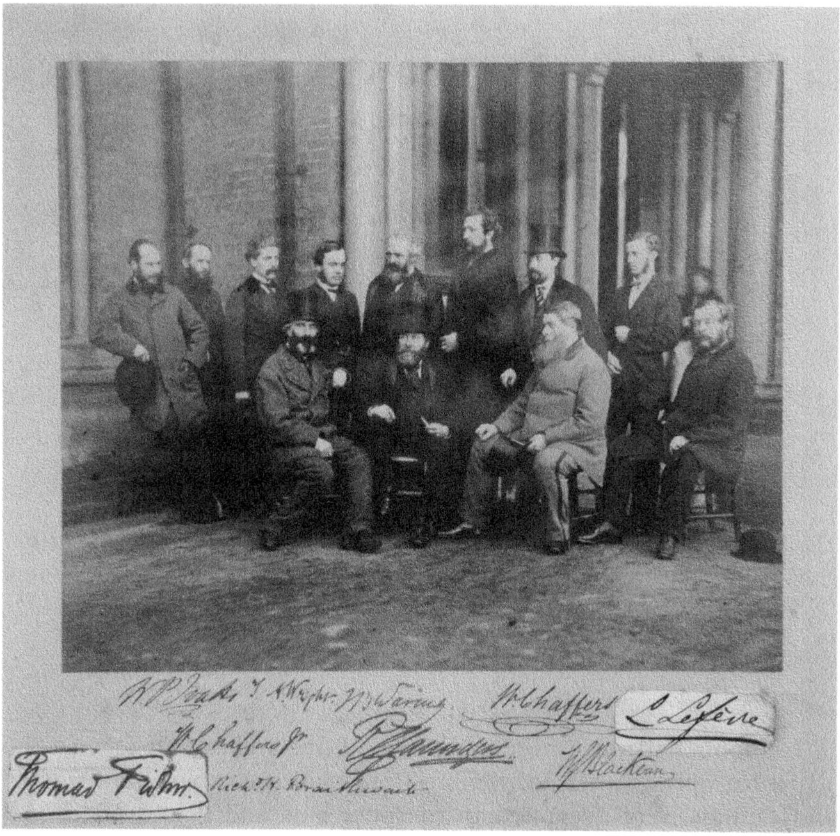

Figure 5.1 Anonymous, *Directors and Officials, Exhibition at New Infirmary Leeds 1868*, 1868, photograph. Leeds Museums and Galleries, UK/Bridgeman Images

Institution and Literary Society in the staging and administration of public exhibitions was secured with a promise of £2,000 for its new building, to be taken from the £20,000 projected profit. Confident of an even greater return than this, the building committee and exhibition committee decided to commission a self-supporting glass and iron structure to cover what Gilbert Scott had intended to be an open courtyard in the centre of the Infirmary.

This new Central Hall was designed by the engineer Rowland Mason Ordish (1824–86), who had been involved in the construction and reconstruction of the Crystal Palace and the roof of St Pancras Station

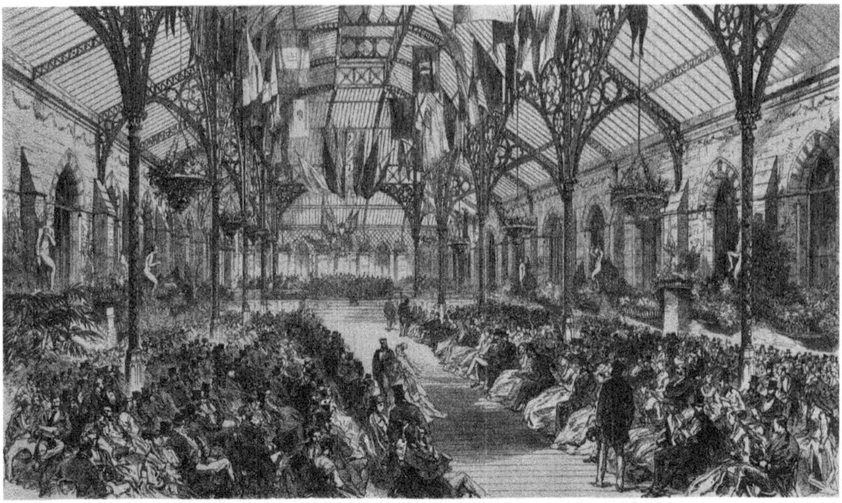

Figure 5.2 Anonymous, *Exhibition of Works of Art at Leeds: The Central Hall*, from *The Illustrated London News*, 28 August 1868, wood engraving. Collection of the author

in London. The twelve iron columns and five arches that formed the structure were painted in 'sombre tints', and the ornamental tracery and capitals were decorated in maroon, blue and gold.[13] Additional temporary decorations were added for the opening ceremony led by the Prince of Wales, including flags, flowers and foliage, surmounted by a bust of Queen Victoria, in symbolic attendance in her role as patron of the exhibition.[14] The Central Hall was built with a capacity of 2,300 people, at an additional cost of £4,000.[15] The structure was intended to benefit the Infirmary after the exhibition closed, as a winter garden for the use of patients. It did go on to function in this way and housed a tennis court until it was demolished in 1911.[16] Fortunately for the exhibition committee—given the risk it had taken by extending the existing deficit on the building—the architectural quotation from the Great Exhibition was not lost on the national press: *The Times*

13 'The National Art Exhibition at Leeds', *The Times*, 16 May 1868, p. 9.

14 'The Prince of Wales at Leeds', *Illustrated London News*, 30 May 1868, p. 541.

15 'Exhibition of Works of Art at Leeds: The Central Hall', *Illustrated London News*, 29 August 1868, p. 196.

16 *The General Infirmary at Leeds: Its Work and Progress, 1767–1911*, p. 12.

described the Central Hall as 'a miniature Crystal Palace', and the professional periodical the *Engineer* praised the structure and published a detailed set of technical illustrations to show its innovative design and execution.[17]

'Planned for the reception of patients rather than of pictures'

The architecture of the Infirmary was the subject of much discussion as to its suitability for the display of painting and sculpture. Designed by one of the most celebrated architects of the Gothic Revival, both the building and exhibition committees were keen to show off what they considered to be 'the most perfect and noble work of its class to be found in Europe' (Figure 5.3).[18] They saw the empty wards as 'peculiarly well suited to the purposes of a Fine-Art Exhibition' because they had been designed to be light, well-ventilated, watertight and fireproof.[19] The permanence and solidity of the structure was emphasised as a point of contrast to the purpose-built but insubstantial spaces that housed other temporary exhibitions, such as the Art Treasures Exhibition in Manchester. While the building received nothing but positive notices for its beauty and projected utility as a hospital, the national press was more sceptical about its appropriation as an art gallery. The *Spectator* noted that the 'galleries have been planned for the reception of patients rather than of pictures' and as a consequence of the large east- and west-facing windows, the level of light was inconsistent, even though the windows had been partially covered to mitigate the worst of the glare.[20] *The Times* described the conditions as 'in the all-important particular of lighting, about as ill adapted as possible for pictures'.[21]

Painting and portraiture

Despite its limitations, the majority of the space in the Infirmary was reserved for the display of painting, which was separated into the following categories: Italian, Spanish and French old masters; German,

[17] 'Leeds Art Treasures Exhibition', *The Times*, 18 May 1868, p. 10; 'The Leeds Exhibition Building', *Engineer*, 22 May 1868, p. 379.
[18] Waring, *Extract from the First Report by J.B. Waring*, p. 4.
[19] Waring, *Extract from the First Report by J.B. Waring*, p. 4.
[20] 'The Arts Exhibition at Leeds', *Spectator*, 23 May 1868, p. 13.
[21] 'Leeds Art Exhibition', *The Times*, 28 May 1868, p. 10.

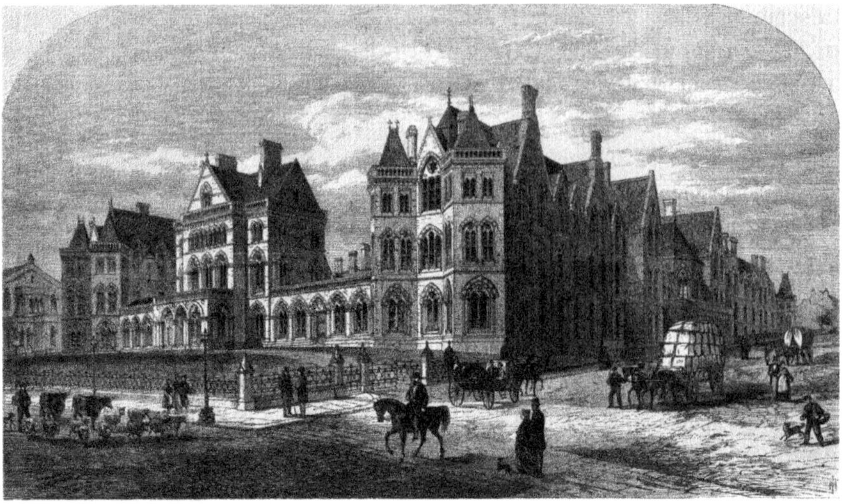

Figure 5.3 Block cut by Mason Jackson, *The New Infirmary at Leeds, Now Occupied by the National Art-Exhibition, Opened by the Prince of Wales on Tuesday*, from *The Illustrated London News*, 23 May 1868, wood engraving. Collection of the author

Flemish and Dutch old masters; deceased English oil painters; living English oil painters; modern foreign oil painters; and English watercolourists, to be visited in that order (Figure 5.4).

The organisers of the exhibition recognised that the status of British painting was not as high as that of European painting. They were keen to argue, however, that what was described as comparative inactivity was not based on an intrinsic deficiency, which they evidenced by pointing towards examples of medieval craftsmanship and the success of contemporary British painting, for which they credited the example set by Hogarth, Reynolds and Gainsborough and more recently by the Pre-Raphaelites, who despite unspecified 'objectionable features', were praised for their 'earnestness of purpose and careful workmanship'.[22]

The relationship between 'foreign' and 'native' painting was held in productive tension, with the highest praise reserved for the work of the Italian old masters. This was tempered by a belief that careful study of these works by British artists would lead them to claim their rightful position of natural heirs to the canon; a familiar narrative since

[22] J.B. Waring, *Baines's Handbook to the Picture Galleries* (Leeds: Baines, 1868), p. 67.

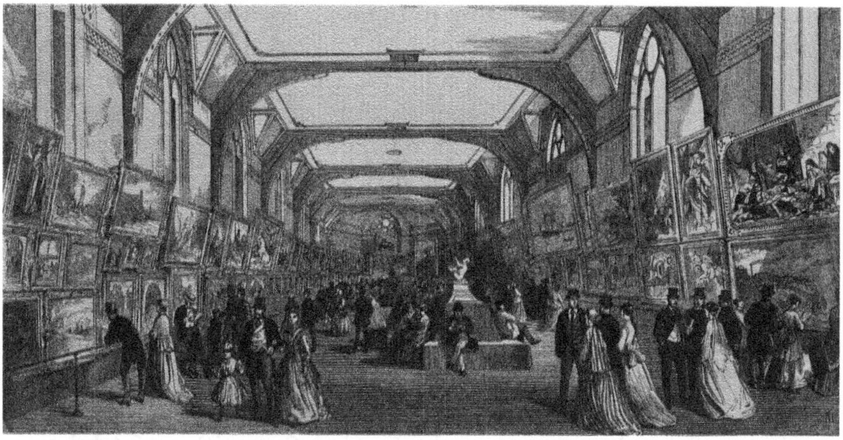

Figure 5.4 Block cut by Mason Jackson, *The Leeds Exhibition: The Gallery of British Living Painters in Oil*, from *The Illustrated London News*, 27 June 1868, wood engraving. Collection of the author

at least the founding of the Royal Academy of Arts a century earlier. The *Illustrated London News* published lengthy and detailed descriptive notices of the 'foreign' portions of the exhibition, often accompanied by wood engravings of works selected for special attention. The periodical foregrounded three sixteenth-century paintings for the benefit of British artists—*The Sultan Mehmet II* attributed to Gentile Bellini, *St Jerome* by Giovanni Girolamo Savoldo and *The Virgin and Child* by Giovanni Battista Cima da Conegliano—which they thought demonstrated more sensitive use of colour than 'those violent lights and unnatural shadows which characterise so many productions of our English school'.[23] It is significant that all three works later entered the collection of the National Gallery as part of the Salting and Layard bequests in 1910 and 1916 respectively. This was the realisation of the plea to wealthy private collectors to, in the words of Waring, 'delight and instruct thousands of their less fortunate fellow countrymen, and to afford the entire nation a source of pure and elevating pleasure, full of instruction to the eye and to the mind'.[24]

In addition to the galleries of painting classified by national school, a portrait gallery that contained 283 oil paintings and 110 engravings

23 'The Art-Exhibition at Leeds', *Illustrated London News*, 20 June 1868, p. 614.
24 Waring, *Extract from the First Report by J.B. Waring*, p. 6.

of 'Yorkshire Worthies' was hung in the long corridors connecting the wards on the ground floor. Collated by Edward Hailstone (1818–90), a local solicitor and prolific book collector, the assembled pictures were intended to represent, in Hailstone's words, 'those persons who, when living, exercised influence in Church and State, whether Prelate or Puritan, Cavalier or Roundhead, and of those who were in any way celebrated in Naval and Military Affairs, Arts, Science, and Literature, or by birth, possessions, or circumstances intimately connected with the county'.[25] It was conceived under the same logic as the relatively recently established National Portrait Gallery in London, in the sense that it was very firmly the subjects that were of primary interest rather than the art objects or their creators. The connection with the National Portrait Gallery was not superficial: one of its founders, Philip Henry Stanhope (1805–75), was part of the London Committee of Advice, as was George Scharf (1820–95), the gallery's first director. The gallery also lent a work from its own collection—a portrait of John Flaxman by George Romney—which was photographed by the London firm Cundall and Fleming and published with 200 of the portraits in oil as part of two large volumes that also contained biographical descriptions of the sitters.[26] Scharf visited the exhibition in late August and made extensive notes and detailed sketches, perhaps with an eye on future acquisitions.[27] In 1919, under Scharf's successor Lionel Cust (1859–1929), the gallery did purchase two sixteenth-century portraits that had been displayed as part of the exhibition in 1868.[28]

A critic from *The Times* concluded that Leeds had equalled—and in some respects exceeded—the display of paintings at the Art Treasures Exhibition in Manchester. They praised the separation of work by living English artists from dead and judged the selection of watercolours to be better, if not greater in number. Leeds was deemed 'decidedly superior in its representation of contemporary continental painting'.[29] The *Illustrated London News* agreed that 'no such important collection

[25] Edward Hailstone, *Portraits of Yorkshire Worthies: Selected from the National Exhibition of Works of Art at Leeds, 1868: With Biographical Notices*, 2 vols. (London: Cundall and Fleming, 1869), p. iv.

[26] Hailstone, *Portraits of Yorkshire Worthies.*

[27] George Scharf, Trustees' sketchbook 14 (22 and 27 August 1868), Heinz Archive and Library, National Portrait Gallery [NPG7/1/3/1/2/15].

[28] Unknown Anglo-Netherlandish artist, *Thomas Wentworth, 1st Baron Wentworth*, oil on panel, 1549–50, National Portrait Gallery [NPG 1851], and unknown artist, *Thomas Wentworth, 2nd Baron Wentworth*, oil on panel, 1568, National Portrait Gallery [NPG 1852].

[29] 'Leeds Art Exhibition', *The Times*, 28 May 1868, p. 10.

of painting by the old and living masters of Europe has been formed in any country' since the Art Treasures Exhibition.[30]

The Museum of Ornamental Art

Curated by the antiquarian William Chaffers (1811–92), this substantial section operated as a miniature South Kensington Museum. Even the display cases appear to have been identical in design to those of its predecessor, the Museum of Manufactures at Marlborough House. Presented in the Chapel, three small adjacent rooms and Gallery J, the Museum of Ornamental Art contained a range of objects diverse in historical period, geographical location, materials and methods: Celtic, Anglo-Saxon, Egyptian, Greek and Roman antiquities; engraved gems; illuminated manuscripts; sculpture; carvings; metalwork; enamel; maiolica; textiles; plate; arms and armour; glass; watches and clocks; bijouterie; English and foreign porcelain and pottery; furniture; miniatures; and a substantial 'oriental' section that represented the decorative arts of China, Japan, Morocco, Algeria, India, Myanmar, Iran and Iraq among other unspecified 'oriental' cultures (Figures 5.5 and 5.6). Loaned by a similar combination of members of the local and national aristocracy, political classes, military figures, clergy and the professions to the rest of the exhibition, the question of how these objects came to be in private collections in Britain is not straightforward to answer. Unlike the display of non-Western objects at Leeds Polytechnic Exhibition of 1845, the Museum of Ornamental Art at the National Exhibition of Works of Art did not explicitly present these objects as curiosities, nor as the spoils of war. Of the twelve examples of 'oriental' arms and armour loaned by Queen Victoria, two flintlock pistols decorated with silver and coral had been presented to George IV as Prince Regent in 1811 and 1819 by the Dey of Algiers, while under French colonial rule.[31] As diplomatic gifts, these objects were invested with additional layers of meaning, but intended in this context to demonstrate excellent craftsmanship. Alongside the separate India Museum to which we will return, these displays, set apart from the works designated as fine art, situated Leeds in relation to the wider British Empire and communicated the town's cultural, commercial and colonial networks and priorities.

30 'Art-Exhibition at Leeds', *Illustrated London News*, 6 June 1868, p. 551.

31 *National Exhibition of Works of Art at Leeds, 1868; Official Catalogue* (Leeds: printed by Edward Baines and Sons, 1868), p. 286.

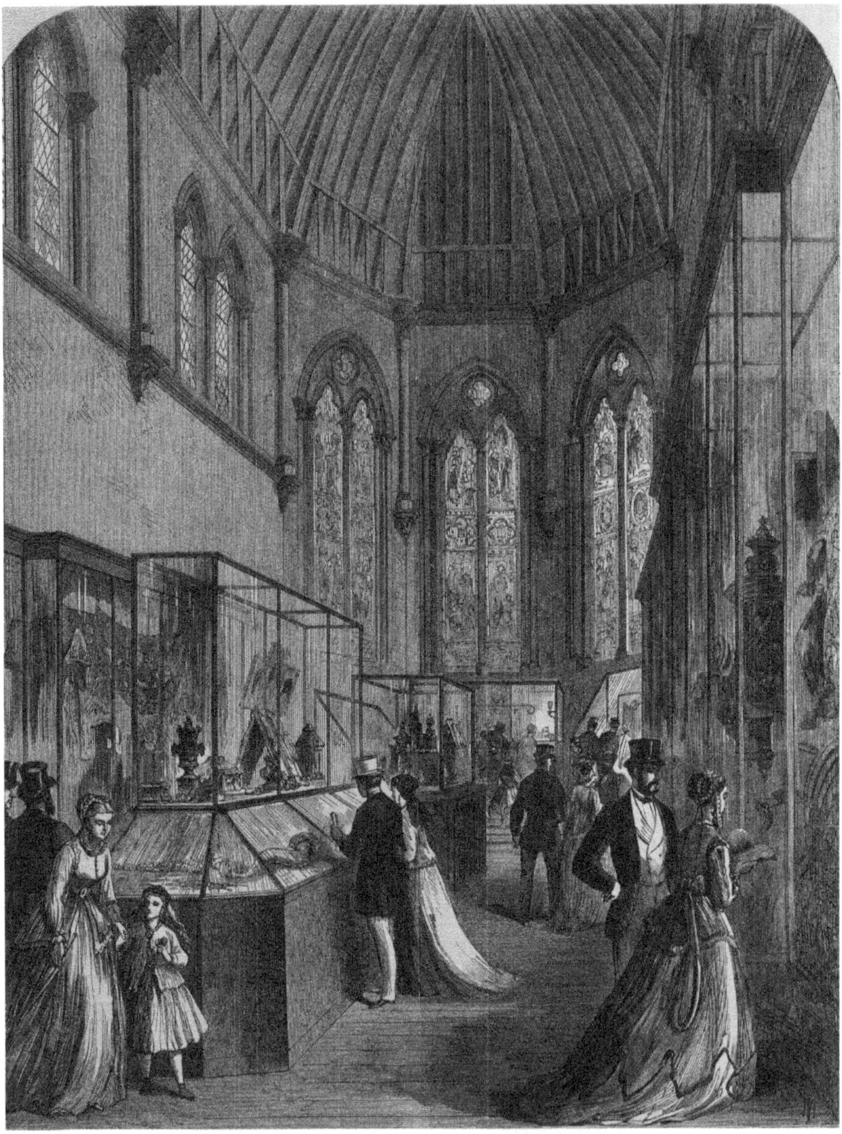

Figure 5.5 Block cut by Mason Jackson, *National Exhibition of Art at Leeds: The Mediæval Museum*, from *The Illustrated London News*, 11 July 1868, wood engraving. Collection of the author

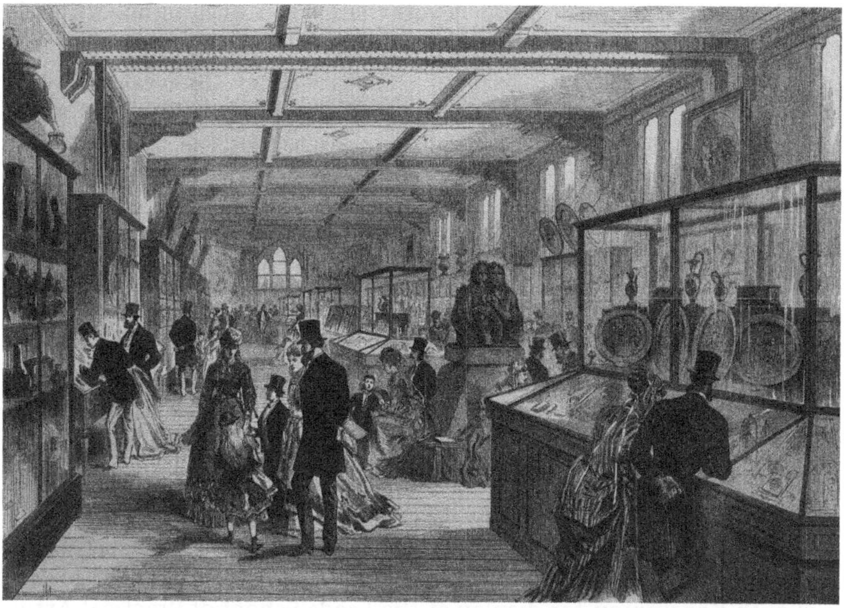

Figure 5.6 Block cut by Mason Jackson, *National Exhibition of Works of Art, Leeds: The Museum of Ornamental Art*, from *The Illustrated London News*, 8 August 1868, wood engraving. Collection of the author

Waring introduced the Museum of Ornamental Art with the following familiar refrain:

> It is hoped that the artizan, especially, will carefully note the more remarkable examples of that class in which he may be personally interested; and carry away with him, not only many valuable ideas, but an increased sense of the estimation in which the best specimens of artistic industry are held by all educated persons, as affording a high idea of the genius of the people who were able to produce them.[32]

It is significant that Waring appeared to address the maker rather than the usual appeal to the employer here. The motivation for the maker to study exemplary objects and improve their own work was located in the

[32] *National Exhibition of Works of Art at Leeds, 1868*, p. 179.

possibility that the result could elevate their status, although Waring did not go so far as to suggest any direct economic benefit to the artisan.

Loans from the South Kensington Museum formed a small part of the Museum of Ornamental Art. A selection of twelve 'important and valuable objects' were displayed in turn over the course of the exhibition. The official catalogue noted many of the purchase prices, with the implication that at least part of their value was located in what had been paid for them, with the two most expensive objects having been bought from the Exposition Universelle in Paris just the previous year: the *Milton Shield* made by Leonard Morel-Ladeuil for Elkington & Co. in 1866, purchased by the South Kensington Museum for £2,000, and a contemporary wooden cabinet made by the firm of Henri-Auguste Fourdinois, which was awarded the Grand Prix and purchased for £2,800. Other objects were presented as notable for their associations with historical figures—Thomas Becket's candlestick, a mirror that belonged to a mistress of Henry II, Diane de Poitiers, Marie Antoinette's harp—and others for their antiquity or local interest.[33] The selection was quite a different proposition to the Circulating Collection loaned by South Kensington for display at the Leeds School of Practical Art in 1855–56 and 1862–63, although the purpose was ostensibly the same.

In addition to the 'Yorkshire Worthies' section, part of the Museum of Ornamental Art was recorded in the form of a publication, illustrated with photographs taken by Cundall and Fleming, of 'objects of archæological interest'. Funded by subscription through the Huddersfield Archæological and Topographical Association and featuring an introduction and description by Chaffers, it formed a partial record of the objects displayed, although the Association expressed its disappointment that it could not have been more extensive: 'That a large and most complete series of illustrations of almost every branch and period of ornamental art, which might have been produced from the specimens assembled at Leeds, was not secured, is much to be regretted.'[34] The *Illustrated London News* did, however, publish wood engravings of some of the objects considered most noteworthy, which included a sword that belonged to John Hampden loaned by Queen Victoria; a Chelsea vase loaned by the Earl of Chesterfield; a silver ewer

33 *National Exhibition of Works of Art at Leeds, 1868*, p. 299.

34 Fairless Barber, ed., *A Series of Photographs of Objects of Archaeological Interest from the National Exhibition of Works of Art at Leeds, 1868: Taken under the Direction of the Council of the Huddersfield Archæological and Topographical Association with Introductory Remarks and Descriptions by William Chaffers* (London: printed by Whittingham and Wilkins, 1869), p. iii.

designed by Flaxman and manufactured by the silversmiths Hunt and Roskell; a silver gilt wine cup; a pair of bellows with decorative carving; the Hunterston brooch; a silver nef in the shape of a ship; an Italian door knocker; a Rockingham vase; a terracotta Bacchus by Clodion; a steel casket; and an Elizabethan silver gilt cup in the shape of a pelican loaned by the politician and antiquarian Sir Stephen Glynne.[35]

Sculpture and the Royal Academy of Arts

The place of sculpture in the exhibition was paradoxically both prominent and marginalised. At the planning stage the committee intended the Central Hall to be 'adorned with sculpture, fountains, and flowers', which might suggest a perceived equivalence between these three things as decoration subservient to the architecture. Writing a year before the opening of the exhibition, Waring set out more detailed plans for the display of sculpture:

> I recommend that it should be kept almost entirely to the ground floor, and that it should form the principal ornament of the Central Hall. Fine pieces of sculpture against the buttresses and a double row forming a central avenue, combined with plants and flowers, will add materially to the beauty of the hall. Groups and Busts will, alternately with other ornamental objects, greatly aid the effective appearance of the main staircases and the corridors of the central court: the screens also of the Picture Galleries, may be rendered very picturesque by a judicious use of smaller works in Sculpture.[36]

With the exception of a bust of Queen Victoria, the sculpture displayed in the Central Hall—and retained by the Infirmary for the winter garden—was a generic set of plaster casts of classical and Renaissance statuary set against the walls. These objects were not identified in the catalogue and were barely mentioned in the otherwise extensive

[35] 'The Leeds Art-Exhibition', *Illustrated London News*, 8 August 1868, p. 127; 'The Leeds National Exhibition of Arts', *Illustrated London News*, 22 August 1868, p. 187; 'The Leeds Exhibition of Works of Art', *Illustrated London News*, 12 September 1868, p. 249; 'National Exhibition of Works of Art at Leeds', *Illustrated London News*, 19 September 1868, p. 272; and 'The Leeds National Exhibition', *Illustrated London News*, 26 September 1868, p. 308.

[36] Waring, *Extract from the First Report by J.B. Waring*, p. 6.

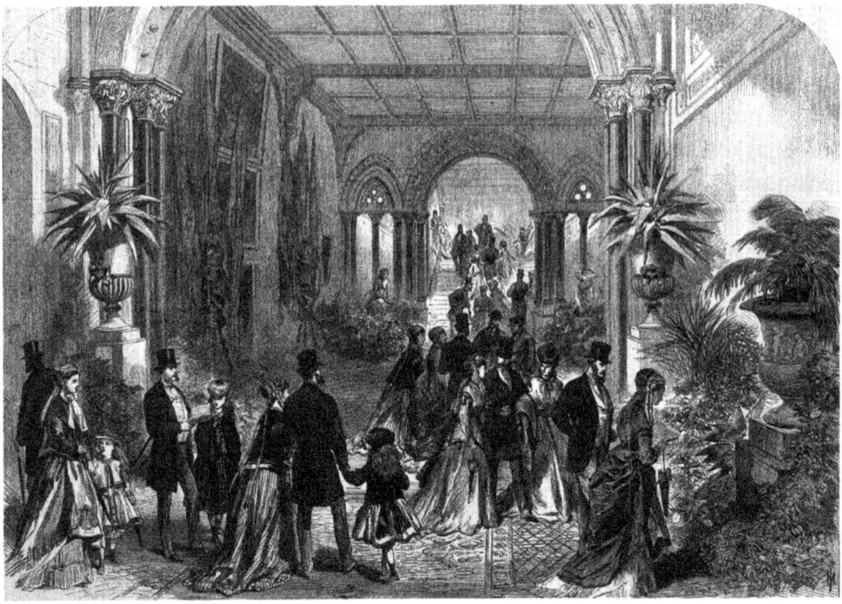

Figure 5.7 Block cut by Mason Jackson, *The Grand Staircase at the Leeds Exhibition of Arts*, from *The Illustrated London News*, 20 June 1868, wood engraving. Collection of the author

commentary, with the *Illustrated London News* simply describing the collection as 'Gods and heroes of old'.[37]

British sculpture was represented by works in marble from the Royal Academy of Arts, the Royal Collections and directly from the studios of artists. The Royal Academy lent Diploma Works by Patrick MacDowell (1799–1870), William Calder Marshall (1813–94), John Henry Foley (1818–74) and John Gibson (1790–1866). Gibson's *Narcissus* greeted visitors at the foot of the principal staircase just beyond the turnstiles, making it one of the first works the visitor encountered (Figure 5.7). It is not clear, however, whether this was considered a prestigious location. The entrance space also contained displays of foliage, armour and paintings, so Gibson's sculpture could be read as part of the general decorative scheme rather than an autonomous work of art selected for particular attention. The catalogue noted that some of the sculptural works were to be found in the corridors, although it did not specify

37 'The Prince of Wales at Leeds', *Illustrated London News*, 30 May 1868, p. 541.

which objects or exactly where they were located, detaching the work from its author and title. Another Diploma Work, Calder Marshall's *Young Satyr*, was displayed incongruously in the gallery devoted to living British painters on the first floor. Positioned on a pedestal above a seating area, the work effectively became an architectural ornament to which visitors turned their backs to take a rest. More deference was shown to a series of marbles of, and loaned by, the British royal family. Representations of Queen Victoria, Prince Albert and their children by Baron Carlo Marochetti (1805–67), Mary Thornycroft (1809–95), Joseph Durham (1814–77) and Matthew Noble (1817–76) were less seen as sculptures and more as surrogates for members of the royal family, whose patronage had been secured but not their presence, with the exception of the ceremonial opening of the exhibition by the Prince of Wales. Noble's busts of the Prince and Princess of Wales were acquired by James Kitson for Leeds Town Hall, where in 1872 they joined the existing sculptures of Victoria and Albert executed by the same artist, installed in 1858 and 1865 respectively.

The India Museum

The India Museum was accommodated in two small rooms adjacent to the western entrance to the exhibition. Although the contents were listed in the official catalogue, these spaces were not included in the 'recommendation to visitors as to the route through the building', and appear to have been treated as ancillary—even the smaller Lace Room was featured in the recommended route.[38] Entered through the corridors hung with pictures of 'Yorkshire Worthies', the India Museum was situated at the opposite side of the Central Hall to the Museum of Ornamental Art and was conceived with similar intentions. The displays included embroidered textiles; carpets and rugs; arms and armour; metalwork in silver, gold, iron, steel, brass, copper and pewter; carvings in stone, ivory, horn and wood; pottery and painted objects.[39] These loans from the India Museum in London were selected by John Forbes Watson (1827–92) in his role as reporter for the Products of India for the India Office and director of the India Museum.[40] Forbes Watson introduced the section with the following endorsement:

[38] See the plan of the building and galleries in *National Exhibition of Works of Art at Leeds, 1868*.

[39] *National Exhibition of Works of Art at Leeds, 1868*, pp. 300–12.

[40] Edward Baines, *National Exhibition of Works of Art, Leeds: Baines's Handbook*

> There is hardly any branch of art applied to industry in which the Indian artisan does not produce works of the very highest excellence—works which, in colour, design, and execution, may serve as models to our own students and manufacturers, and which are, indeed, in many respects, unapproached by the highest efforts of European skill.[41]

The message was consistent with the ways in which Indian manufactures had been instrumentalised through public display, particularly since the Great Exhibition of 1851. Deeply embedded in the colonial project, in 1860 Forbes Watson began to collate textile samples from the extensive acquisitions made for the Exposition Universelle in Paris in 1855, to form *The Textile Manufactures and the Costumes of the People of India*. Initially published in 1866 as a set of eighteen volumes in an edition of twenty, each edition included 700 samples of Indian textiles with accompanying texts. Of the twenty editions, thirteen were distributed to manufacturing centres in England, Scotland and Ireland and seven to India. These sample books were intended to operate with the same economic logic as the broader collection: to present examples to inform and improve British manufactures. Where the process was more insidious, however, was that these mechanised manufactures were to be exported back to India in order to exploit the much larger market for cheap printed cottons, which had a ruinous effect on parts of the domestic handloom industry.[42]

Conclusion

The exhibition attracted 570,000 visitors over the 143 days it was open, including 150,000 'excursionists' who arrived by train, with tickets sold by agents in Bradford, Wakefield, Halifax, Huddersfield and Manchester.[43] Despite these impressive figures it generated around £17,000 less in profit than the committee had anticipated, leaving a shortfall of £2,000 for the exhibition and £20,000 for the Infirmary building after heavy expenses of around £52,000.[44] The committee blamed the deficit on a

to the Picture Galleries; a reprint, with additions, of critical notices published in the Leeds Mercury (Leeds: Edward Baines and Sons, 1868), p. vii.

41 *National Exhibition of Works of Art at Leeds, 1868*, p. 300.

42 See Deborah Swallow, 'The India Museum and the British-Indian Textile Trade in the Late Nineteenth Century', *Textile History*, 30:1 (1999), 29–45.

43 'Close of the Leeds Exhibition', *The Times*, 2 November 1868, p. 5.

44 'The Leeds Exhibition and the New Infirmary', *Bradford Observer*, 23 March 1869, p. 4.

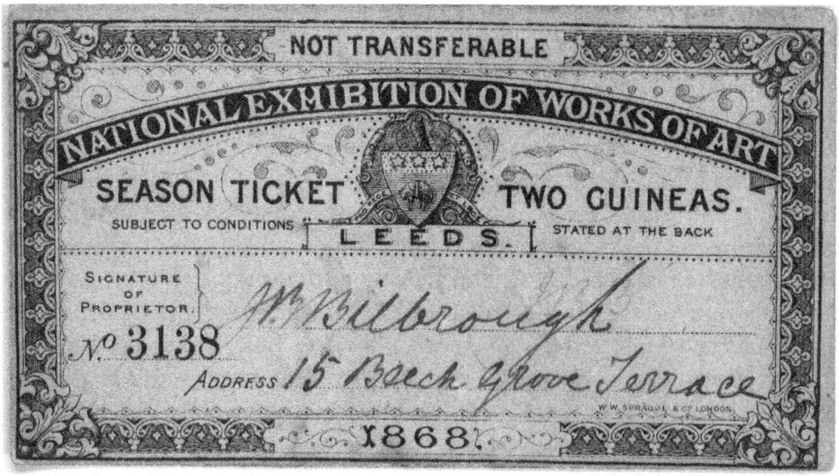

Figure 5.8 Season ticket for the National Exhibition of Works of Art, Leeds, 1868, card. Leeds Museums and Galleries, UK/Bridgeman Images

lack of uptake of the expensive season tickets, priced at five guineas for a first-class ticket and two guineas for a second-class ticket (Figure 5.8). Only 80 of the 187 guarantors had purchased the more expensive season ticket.[45] Almost 80% of visitors paid the day rate of between one and five shillings at the turnstile.[46] Just as the embarrassment of the financial failure of the Leeds Academy of Arts in 1855 had quickly resulted in a public subscription to meet its obligations, by March 1869 £10,000 had been subscribed towards the debt on the Infirmary building.[47]

It was perhaps this inability to meet its basic economic objectives that contributed to the exhibition's general absence from the cultural history of Leeds and from histories of British exhibitions more broadly. Its significance, however, is not best measured by its balance sheet but by its contribution to a genuinely national dialogue about the past, present and future state of British art. The *Illustrated London News* was right when it speculated that 'this is probably the last exhibition of

45 'The Leeds Exhibition and the New Infirmary', p. 4.

46 'Country News', *Illustrated London News*, 7 November 1868, p. 439. The ticket price for the first four days of the Exhibition was 5s. For the rest of May it was reduced to 2s. 6d. From 1 June, on Wednesdays and Fridays the ticket price was 2s. 6d. and on all other days it was further reduced to 1s. See 'The Royal Visit to Leeds', *York Herald*, 14 May 1875, pp. 6–7.

47 'The Leeds Exhibition and the New Infirmary', p. 4.

anything like its value and extent which can take place in this country for many years, if ever in our day'.[48] The Leeds Exhibition operated as a staging post for some of the most significant British painting, sculpture and decorative arts, between the private collections from which they were mostly drawn and the regional, national and international public collections many of these objects would later enter. *The Times* concluded that 'the Exhibition itself is one of the finest displays of the kind that has ever been got up in England out of London. It is not better than that at Manchester in 1857, but it is in its way quite as good.'[49] The *Spectator* decided that, 'on the whole, this Leeds Exhibition seems to be the finest as well as the most extensive Art display ever held in the provinces'.[50] Its correspondent went on to offer the prescient assessment that 'even if this Exhibition proves financially a failure, it may yet do much good'.[51]

The exhibition demonstrated that there was an appetite and an audience for the fine arts in Leeds and the surrounding areas. It served to bolster a sense of civic and cultural confidence that had been so shaken by the collapse of the Leeds Academy of Arts in the previous decade. Although it contained far more than paintings, prints, drawings and sculptures, the National Exhibition of Works of Art was not a polytechnic display on the model of those staged in Leeds in the 1830s and 1840s. It showed that the public—both local and national—would support exhibitions of art without industry, although the next major exhibition in Leeds returned to the proven combination of arts alongside manufactures in order to generate a more secure profit. Although its objectives were not realised immediately, the National Exhibition of Works of Art at Leeds in 1868 ultimately resulted in the successful opening of the new Leeds General Infirmary as a functional hospital on 19 May 1869, national recognition for the cultural status of the town, and a powerful incentive to continue the work of campaigning for a permanent art gallery.

48 'Fine Arts', *Illustrated London News*, 19 September 1868, p. 273.
49 'Leeds Art Treasures Exhibition', *The Times*, 18 May 1868, p. 10.
50 'The Arts Exhibition at Leeds', *Spectator*, 23 May 1868, p. 13.
51 'The Arts Exhibition at Leeds', p. 13.

6

The Yorkshire Exhibition of Arts and Manufactures, 1875

Six years had passed since the National Exhibition of Works of Art in 1868 and Leeds was yet to stage another major public exhibition. In the summer of 1874—thirty-one years after its last comparable polytechnic display in 1843—the Leeds Mechanics' Institution and Literary Society announced its intention to organise the Yorkshire Exhibition of Arts and Manufactures, set to open on 1 May 1875.[1] It began with an economic argument for another exhibition, claiming that since the First Public Exhibition of 1839, 'in almost every case there has been a handsome surplus', and further, 'the exceptions were caused by conditions that ought not to have existed'.[2] These unspecified conditions must have been considered surmountable by the new committee of management established for the purpose of organising the exhibition. Led by the Mayor of Leeds, Henry Rowland Marsden (1823–76), the committee stated its objective: 'to promote the fine arts and arts and science as applied to manufactures' and, once again, to generate sufficient profits to liquidate the debt remaining against the Leeds Institute building.[3] The committee published a prospectus to promote the Leeds Mechanics' Institution and Literary Society, which, they emphasised, 'does not belong to, nor does it seek to serve, any party, sect, or creed. It belongs to and seeks to serve all classes alike, in the advancement of their mental and social well-being.'[4] The committee described the nature of the Institution's public service as the provision of evening classes; girls' and boys' schools; the School of Art; the School of Science; a winter

1 'Yorkshire Exhibition of Arts and Manufactures', *Leeds Mercury*, 13 June 1874, p. 12.

2 'Yorkshire Exhibition of Arts and Manufactures', p. 12.

3 'Yorkshire Exhibition of Arts and Manufactures', p. 12.

4 *Prospectus of the Yorkshire Exhibition of Arts & Manufactures, at Leeds*, 2nd edn (Leeds: Fred R. Spark & Co.), unpaginated.

lecture series; a library and a reading room. The prospectus further explained the predicament of the Institution and its 'incubus of debt':

> In order to provide convenience for its work, and proper accommodation for its classes, the present noble building was erected, which, with fittings and means for extension, has, notwithstanding the very liberal subscriptions of the wealthy and middle classes, a debt of nearly £10,000 left upon it. The interest of this large sum having to be met from the income of the Institution and Classes cripples the Committee, and prevents their doing that which they would gladly do to advance the efficiency of the various departments. The Committee exceedingly regret their want of means to provide Pictures and other objects of Art for the Gallery they already have in the building, and which, if provided, they believe would be a benefit to the pupils of the School of Art, the public, and the trade of the district.[5]

The proposed exhibition was an emphatic return to the presentation of industry alongside the fine and decorative arts. 'Designers, inventors, manufacturers, or producers of articles' were explicitly encouraged to exhibit their own work and offer it for sale.[6] The committee stated that the objects would be classified into two departments: Arts and Manufactures and Fine Arts, with the latter category expected to include 'Paintings in oil and water-colours, and drawings. Sculptures, models, die sinking, and intaglios. Etchings and engravings. Antiquities [and] Photography'.[7] Unlike the polytechnic exhibitions of the late 1830s and early 1840s that were organised in less than six months—and often even less than that—the Mechanics' Institution announced its intentions almost a year in advance of the planned opening date, which contributed to the sense that the committee were determined to avoid the pitfalls of their predecessors. Their prudence extended to the preparation of contingencies in case the exhibition failed to return a profit:

> In order to secure the confidence of exhibitors, and to provide against a possible, but not probable, deficiency the General Committee of Management will raise, immediately, a guarantee fund of not less than ten thousand pounds. It has also been thought

5 *Prospectus of the Yorkshire Exhibition of Arts & Manufactures*, unpaginated.
6 'Yorkshire Exhibition of Arts and Manufactures', p. 12.
7 'Yorkshire Exhibition of Arts and Manufactures', p. 12.

> advisable, in order to meet the wishes of gentlemen who do not approve of becoming guarantors, but who desire to aid important public objects, to organise a special donation fund. Donations to the fund will be placed to a separate account, and kept distinct from the exhibition accounts. In the case of a deficiency, one-half of this fund will be applied to its liquidation.

Securing the confidence of exhibitors was crucial to the success of the venture, and these early reassurances were coupled with the announcement of 'Substantial Money Prizes ... to Exhibitors in the Operatives Department', to be awarded by specialist jurors.[8] Potential exhibitors were invited to apply for space by 1 March 1875 by writing to the general manager of the Leeds Mechanics' Institution, J.O. Dayson.[9] Composed of both individuals and private companies, 238 guarantors pledged a total of £12,100. Led by Marsden's contribution of £500, Beckett & Co. pledged £400; four further individuals and companies £250 each; 45 at £100; 67 at £50; 69 at £25; 19 at £20; 17 at £10; and 15 at £5.[10] As a guarantee fund established to insulate against the possibility of financial loss, the plan was not to have to call on the guarantors by ensuring that the exhibition at least broke even. Given that the exhibition cost £26,782. 5s. 6d. to stage, it is important to ask why a subscription was not raised to alleviate the Mechanics' Institution's debt instead.[11] Although the guarantors no doubt hoped not to have to pay out their pledges, they were presumably able to do so should the situation have arisen. As such, the exhibition, and their public support of it, provided tangible and intangible benefits above the simple liquidation of the debt against the Mechanics' Institute building.

By December 1874 the opening date had been revised to 6 May 1875—a familiar delay, although the announcement was much earlier than its predecessors, giving five months' notice of the change.[12] In January 1875 a new 'Sanitary Section' was announced and the exhibition

[8] 'Yorkshire Exhibition of Arts and Manufactures', *Sheffield Independent*, 12 December 1874, p. 5

[9] 'Yorkshire Exhibition of Arts and Manufactures', *Leeds Times*, 6 February 1875, p. 4.

[10] *Prospectus of the Yorkshire Exhibition of Arts & Manufactures, at Leeds*, unpaginated.

[11] 'Result of the Recent Exhibition of Arts and Manufactures in Leeds', *Sheffield Independent*, 30 November 1876, p. 6.

[12] 'Yorkshire Exhibition of Arts and Manufactures', *Sheffield Independent*, 26 December 1874, p. 5.

was advertised to run until the end of September 1875.[13] The new year also brought the significant news that Queen Victoria had agreed to be the patron.[14] As we have seen, royal patronage was highly prized and, once granted, extensively advertised as a means of both promoting the exhibition and providing a reassuring 'seal of approval' to both prospective exhibitors and visitors. The historian Simon Gunn advanced the argument that 'the public prestige of the monarchy was reinvigorated from the 1870s and used, with considerable effect, to lend ceremonial dignity to the British state'.[15] This rejuvenation occurred at a local level in Leeds, where it was hoped that another royal visit would confer 'ceremonial dignity' on the town and its new exhibition.

The advanced planning extended to the sale of season tickets, which were made available for sale from 10 March 1875. The cost of a season ticket that would admit the holder to the opening ceremony and three more 'Special Ceremonials' was £2 2s.[16] As the most expensive ticket, its cost matched the price of the second-class season ticket to the National Exhibition of Works of Art in 1868 and avoided repeating the mistake of the five guinea first class season tickets that were generally blamed for the financial failure of the previous exhibition.[17] A number of other tickets were available, including a half-price season ticket at £1 1s. which excluded the ceremonial days.[18] The cost of single admission tickets was stratified too, with a 2s. 6d. ticket for Fridays and a 1s. ticket for all other days, and children under 12 at half price.[19] Tickets were available to purchase from the Mechanics' Institute, from Mr Inchbold on Bond Street (Beck and Inchbold was a manufacturing stationers at 33 Bond

13 'Yorkshire Exhibition of Arts and Manufactures', *Sheffield Independent*, 12 January 1875, p. 1.
14 'The Queen and the Yorkshire Exhibition of Arts and Manufactures', *Sheffield Daily Telegraph*, 30 January 1875, p. 3.
15 Simon Gunn, 'Ritual and Civic Culture in the English Industrial City, c. 1835–1914', in *Urban Governance: Britain and Beyond since 1750*, ed. by Robert J. Morris and Richard H. Trainor (Aldershot: Ashgate, 2000), pp. 226–41 (p. 226).
16 'Yorkshire Exhibition of Arts and Manufactures', *Bradford Observer*, 6 March 1875, p. 1.
17 'The Royal Visit to Leeds', *York Herald*, 14 May 1875, pp. 6–7.
18 Tickets for the Inaugural Ceremony only were advertised on 5 May for one guinea (£1 1s.) in 'Yorkshire Exhibition of Arts and Manufactures, Leeds, 1875', *Bradford Observer*, 5 May 1875, p. 1.
19 'Prospectus of the Yorkshire Exhibition of Arts & Manufactures, at Leeds', unpaginated.

Street) and Mr Archibald Ramsden on Park Row (Ramsden was a musical entrepreneur with a purpose-built saloon at 12 Park Row).[20]

'Loyalty and Official Robes'

In April 1875 it was announced that the exhibition would be inaugurated by the Duke of Edinburgh on 13 May 1875.[21] The news of a royal visit quickly generated opportunistic advertisements for flags, banners, illuminations and floral arrangement intended to decorate the streets and buildings.[22] The Town Council swung into action too with a special meeting to discuss the invitation for councillors to attend the opening in an official capacity. The resulting discussion was more contentious than might have been expected, when Alderman Kelsall expressed the following opinion about how they should appear on this public occasion, published in an article in the national newspaper *The Times*:

> He hoped every member of the Council would appear on the occasion in question in his official robe, because he knew it was the wish of the Queen that such should be the case. He knew this because when the late Alderman Sir Peter Fairbairn, Alderman George, and himself were presented to Her Majesty in anticipation of her visit to Leeds in 1858, along with the late Prince Consort, and two of the Royal Princesses, she expressly stated most condescendingly that she desired that the members of the Corporation should appear before her at Leeds in their official costumes and not in the ordinary Court dresses.[23]

While Alderman Blackburn seconded the motion, dissent emerged among the group and it was reported that Councillor Scarr 'thought some of the Council were thinking more of the robes than of the man'. Scarr continued:

> There were people who wished to pay all deference to Her Majesty and the Royal Family under any circumstances, but who, in order to

[20] 'Yorkshire Exhibition of Arts and Manufactures at Leeds', *Leeds Times*, 24 April 1875, p. 4.
[21] 'Yorkshire Exhibition of Arts and Manufactures at Leeds', p. 4.
[22] 'The Royal Visit to Leeds', *Leeds Times*, 24 April 1875, p. 4; 'Yorkshire Exhibition at Leeds', *Leeds Times*, 24 April 1875, p. 4.
[23] 'Loyalty and Official Robes', *The Times*, 26 April 1875, p. 14.

> do so, did not think it necessary to don a mere robe as the emblem of their allegiance. Sometimes robes were donned in order to show allegiance when no allegiance at all was intended, and though he was always ready to show his respect for the Queen, he repeated that he should never accept the dictation of any one as to the utility or custom of wearing a purple robe.[24]

Scarr was joined in his opposition by Mr Barker, who objected to the expense of the 'puppet show' and the imposed deference to monarchy. The mayor reassured his colleague that he would meet the costs himself as president of the exhibition committee.[25] Scarr followed through with his objection and his protest was noted on the day of the opening by the *York Herald*:

> Some notice was likewise bestowed on the appearance in one of the Corporate carriages of Mr. Scarr, a member of the municipal council, who had with considerable ostentation announced his intention of expressing his sense of opposition to what he would call baubles and man millinery by refusing to follow the official costume donned by his colleagues. He as was good as his word. He appeared in the everyday dress of a civilian.[26]

Scarr's position was consistent with his politics. Archibald Witham Scarr (1827–1904) was a working-class market trader by background, who had successfully expanded the family business and entered local politics. A supporter of the franchise for working men, temperance and Smilesian self-help, Scarr rejected what he saw as excessive spending of ratepayers' money.[27]

Dissenting voices were, however, overwhelmed by the weight of enthusiasm generated in the periodical press, both local and national. Even the particulars of the exhibition itself were subsumed underneath the excitement of a royal visit. The *Leeds Mercury* set the visit in context by providing readers with an account of the history of the monarchy's relationship with Leeds over the previous forty years, from Queen Victoria's first visit as Princess Alexandrina Victoria of Kent in 1835 along with her mother, to the first visit after her accession to open the

24 'Loyalty and Official Robes', p. 14.

25 'Loyalty and Official Robes', p. 14.

26 'The Royal Visit to Leeds', *York Herald*, 14 May 1875, pp. 6–7.

27 See Herbert Yorke, *A Mayor of the Masses: History and Anecdotes of Archibald Witham Scarr* (Leeds: Morrison and Roberts, 1904).

new Leeds Town Hall in 1858, followed by the opening of the National Exhibition of Works of Art by the Prince of Wales in 1868, and most recently, the opening of Roundhay Park by Prince Arthur in 1872.[28]

The Coloured Cloth Hall

The scale of the exhibition and its royal patronage attracted commentary from newspapers based in London, and their reviews were generally encouraging, although they tended to focus on the good behaviour of the crowds on the opening day and the reception given to the Duke of Edinburgh. As Gunn noted in his research on ritual and civic culture in this context, these occasions were treated as 'expressions of order', which were interpreted as 'an index of civility'.[29] The *Daily News* reported its impression of the exhibition's site at the Coloured Cloth Hall:

> a large and somewhat ugly building, situate [*sic*] near the railway stations in Wellington-street, but one which is admirably adapted for the purpose of a fine art gallery, owing to the peculiar style of the erection and the admirable manner in which it is lighted—plenty of light being a sine qua non in a building where manufacturers have to dispose of delicately hued and shaded cloths to merchants.[30]

The appropriation of commercial spaces for the purposes of public exhibitions reinforced the relationship between culture, capital and consumption in mid-nineteenth-century Leeds. The Covered Market, Coloured and White Cloth Halls and Stock Exchange were used to provide architectural scale, urban centrality and signifiers of civic culture and commercial prosperity.[31] With reference to Birmingham, Manchester and Leeds, Gunn argued that nineteenth-century urban centres 'developed as regional as well as industrial capitals. This role was

[28] 'The Royal Visit to Leeds', *Leeds Mercury*, 13 May 1875, p. 7.

[29] Gunn, 'Ritual and Civic Culture in the English Industrial City, *c*. 1835–1914', p. 233.

[30] 'Yorkshire Exhibition of Arts and Manufactures. The Duke of Edinburgh's Visit to Leeds', *Daily News*, 13 May 1875, p. 2.

[31] Kevin Grady has noted that the Coloured Cloth Hall could accommodate 1,770 stalls within the interior and 20,000 people in the courtyard. See Kevin Grady, 'Commercial, Marketing and Retail Amenities, 1700–1914', in *A History of Modern Leeds*, ed. by Derek Fraser (Manchester: Manchester University Press, 1980), pp. 177–99 (p. 180).

symbolised by the exchanges and cloth halls that formed the meeting place and hub for the industries that colonised the larger manufacturing districts.'[32] The Cloth Halls had been used to stage exhibitions and other public spectacles since the late eighteenth century; an early example in Leeds was an 'Aerial Excursion' by the Italian aeronaut Vincenzo Lunardi (1754–1806), who departed from the yard of the White Cloth Hall by hot air balloon in 1786.[33] The centrality of the cloth halls to both the spatial and economic composition of the town made them ideal for temporary public events that sought to align commercial prosperity with cultural prestige. The historian Kevin Grady has argued that 'the importance of the halls cannot be overstated: contemporaries thought them the foundation of Leeds's commercial success in the eighteenth and nineteenth centuries'.[34]

Built in 1758, the Coloured Cloth Hall could accommodate 1,770 stalls and up to 20,000 people in the courtyard of its massive quadrangular structure.[35] Returning to the comment in the *Daily News* quoted above, that the writer considered the building to be 'somewhat ugly' was perhaps connected to its age and purpose. In use for over a century by 1875, it no longer communicated progress and innovation, instead operating as a relic of the earliest days of a textile industry that had been transformed by mechanisation and the factory system. It is important to note that the criticism of the external appearance of the building was not limited to this one notice in the metropolitan press; local commentary found fault with the exterior too, while the alterations to the interior were generally met with approval. This example from the *Leeds Mercury* contains both sentiments:

> Few visitors, after once entering the Exhibition, would ever suppose from the scene around them that they were actually within the historical precincts of the Old Cloth Hall. No transformation would have been more complete and surprising. Whilst enjoying, however, all the fresh interest of such a change, let us not altogether forget that the place has a story of its own, recalling bygone days when such an enterprise would have been thought impossible. The original "hall," forming a kind of shell for the Exhibition, is

32 Simon Gunn, *The Public Culture of the Victorian Middle Class: Ritual and Authority and the English Industrial City 1840–1914* (Manchester: Manchester University Press, 2000), p. 12.

33 David Thornton, *The Story of Leeds* (Stroud: The History Press, 2013), unpaginated.

34 Grady, 'Commercial, Marketing and Retail Amenities, 1700–1914', p. 180.

35 Grady, 'Commercial, Marketing and Retail Amenities, 1700–1914', p. 180.

> an unsightly pile of brick buildings in a quadrangle form... The exterior is not yet at all prepossessing, but a very spacious and suitable interior has been secured.[36]

Despite its style of architecture having fallen out of fashion, the proximity of the Coloured Cloth Hall to the Wellington and New Stations for train travel (cheap tickets were made available especially for the exhibition) and the sheer size and flexibility of its internal and external spaces were substantial points in its favour.[37] It is worth noting that the building was demolished in 1890, only fifteen years after the exhibition, which suggests that it was already on its way to obsolescence by 1875. In its extended review the *Leeds Mercury* noted that 'the Cloth Hall has itself been to a large extent superseded by the erection of large warehouses for all the principal firms in town, as well as for manufacturers in the neighbourhood'.[38] The scale of the courtyard did, however, provide a way for the organisers to bring the space up to date and connect their ambitions with the recent history of temporary exhibition architecture. Much like the decision made for the National Exhibition of Works of Art in 1868, the exhibition committee agreed to cover the courtyard with a temporary structure. This central space was designated the 'Edinburgh Hall' to mark the exhibition's inauguration by the Duke of Edinburgh.[39]

The Edinburgh Hall

The temporary Edinburgh Hall was built using the 'flitch' system, which had been deployed by the architect and engineer Francis Fowke (1823–65) for the International Exhibition of 1862. The advantage of this method of construction was in its use of standardised lengths and shapes of wood, which could be reused after the structure had been dismantled. More importantly from the perspective of the committee and their obligation to return a profit, it meant that the materials could be sold on once they had served their purpose.[40] This intent was confirmed in the exhibition's prospectus, which offered both the Edinburgh Hall and the Machinery

36 'The Yorkshire Exhibition', *Leeds Mercury*, 14 May 1875, pp. 7–8 (p. 7).

37 'Yorkshire Exhibition of Arts and Manufactures at Leeds', *Leeds Times*, 10 April 1875, p. 4.

38 'The Yorkshire Exhibition', p. 7.

39 'Yorkshire Exhibition of Arts and Manufactures at Leeds', *Leeds Times*, 10 April 1875, p. 4.

40 The local building firm of Whiteley & Co. were the contractors tasked

Annex for sale, with the buyer responsible for their removal. No price was listed but offers were to be made before 20 June 1875. The result of this advertisement appears not to have been recorded.[41] While continuing to reference the architecture of metropolitan temporary exhibitions, the construction of the Edinburgh Hall avoided the onerous cost of a bespoke iron and glass structure of the kind that so burdened the budget of the National Exhibition of Works of Art in 1868. The Edinburgh Hall was a relatively early commission for the architect William Bakewell (1839–1925). Bakewell became a member of the Freemasons in December 1869 at the age of 31 and joined the Lodge of Fidelity, which met at the Old George Inn on Briggate in the centre of Leeds.[42] This is likely to have coincided with his move to the town and provided him with a professional network through which to secure commissions. Born in Hampstead in north London, Bakewell lodged with a joiner and his family near Durham at the time of the 1861 Census, where he was recorded as a 22-year-old architect.[43] By the time of the next census ten years later, he had moved to Leeds, where he established his architectural practice and became one of the leading designers of late Victorian and early Edwardian architecture in the borough.[44]

The Edinburgh Hall continued the practice of designating a large central space, ornamented with a limited number of impressive exhibits, as a setting for ceremonies, concerts and other social occasions. The Saloon of the Albion Street Music Hall operated in this way during the exhibitions of 1839, 1843 and 1845, as did the Central Hall of the National Exhibition of Works of Art in 1868. The Coloured Cloth Hall was the largest of these spaces at 56 metres long, 30 metres wide and 23 metres high, with 2,000 seated audience members, 400 vocalists and 50 instrumentalists present at the opening ceremony.[45] A 'Grand Orchestral

with the construction of the Edinburgh Hall. Boothman and Broomhead constructed the Machinery Annex.

41 *Prospectus of the Yorkshire Exhibition of Arts & Manufactures*, unpaginated.

42 Library and Museum of Freemasonry, London, Membership Registers 1751–1921, Lodge Number 289, Folio Number 111.

43 Census Returns of England and Wales, 1861, Class: RG 9; Piece: 3738; Folio: 77; Page: 9; GSU roll: 543180.

44 Census Returns of England and Wales, 1871, Class: RG10; Piece: 4565; Folio: 147; Page: 32; GSU roll: 847142. Bakewell would go on to design the Coliseum Theatre (1885), extensions to Tower Works (1899), Athenæum House (1890), London and Midland Bank (1892), Holbeck Public Library (1901), City Square (1903) and Pearl Chambers (1911).

45 'Yorkshire Exhibition of Arts and Manufactures at Leeds', *Leeds Times*, 10 April 1875, p. 4.

Organ' was installed in the space, loaned by Gray and Davison, the London-based manufacturers responsible for the organ at Leeds Town Hall. The organ was originally built for St James's Hall in London, a concert hall marginally smaller than the Edinburgh Hall.[46] In addition to its musical contribution, the organ was also an object through which the town could demonstrate its expertise in engineering, as the air flow through the 2,439 pipes was provided by Hathorn, Davis and Davy's hydraulic engines.[47] As a 'richly decorated instrument, it combined the decorative arts, music and industry operating in harmony, with the visual spectacle and huge sound lending an ecclesiastical resonance to the space.[48]

Music and ceremony were not the only functions of the Edinburgh Hall. The organ was joined by a six-pound field gun from Greenwood and Batley of Armley Works and a 'Grand Park Fountain' by Handyside & Co. of Derby.[49] The impressive fountain was pictured as part of a wood engraving published in the *Illustrated London News*, which showed a highly ornamented structure surmounted by a draped female figure.[50] It was reported that the fountain was 'capable of throwing a jet of water to the roof of the building'—a height of 75 feet, or almost 23 metres.[51] Alongside these demonstrations of the foundry's capacity to produce objects of destruction and decoration, the Hall included aquaria filled with live fish and aquatic plants—one of which was 24 metres long—and an aviary 'with richly plumed birds'.[52] The inclusion of live specimens marked a return—on an expanded scale—to the variety of displays last seen in Leeds as part of the polytechnic exhibitions of the late 1830s and early 1840s. In addition to differing in scale, these displays were more explicitly framed as scientific than their earlier counterparts. One section of the display was dedicated to 'piscinculture'—mostly methods

46 *Prospectus of the Yorkshire Exhibition of Arts & Manufactures*, unpaginated. The organ had 70 stops and was reportedly valued at £1,500.

47 *Prospectus of the Yorkshire Exhibition of Arts & Manufactures*, unpaginated. The organ was the first object listed in the catalogue. See *Yorkshire Exhibition of Arts and Manufactures, Leeds, 1875, Patron: Her Majesty the Queen. Official Catalogue* (Leeds: Inchbold and Beck, 1875), p. 29.

48 'Yorkshire Exhibition of Arts and Manufactures. The Duke of Edinburgh's Visit to Leeds', *Daily News*, 13 May 1875, p. 2.

49 See *Yorkshire Exhibition of Arts and Manufactures, Leeds, 1875*, p. 29.

50 'The Duke of Edinburgh Opening the Yorkshire Exhibition of Arts and Manufactures at Leeds', *Illustrated London News*, 22 May 1875, pp. 496–97.

51 'Yorkshire Exhibition of Arts and Manufactures. The Duke of Edinburgh's Visit to Leeds', *Daily News*, 13 May 1875, p. 2.

52 *Prospectus of the Yorkshire Exhibition of Arts & Manufactures*, unpaginated.

of farming salmon—and the second to freshwater animals and plants. The living specimens were supplemented by a table display of painted casts of fish by local sculptor Matthew Taylor (1837–89), described as having been 'coloured so perfectly, and the form of each fish is so well retained, that this mode must be esteemed preferable, for museum purposes, to either stuffing or preservation of the real fish in spirits'.[53]

The opening ceremony

As a result of having secured a royal visit to inaugurate the exhibition, the opening ceremony dominated discussion in the local and national periodical press (Figure 6.1). Readers were primed for the event through detailed accounts of the itinerary in the *Daily News* and the *Leeds Mercury*, which began with the Duke of Edinburgh's train journey from London King's Cross to Leeds, the train arriving at the Great Northern Station on Wellington Street at 12.35 p.m.[54] As part of the extensive coverage published in the *York Herald*, its reporter described the bright spring day that greeted the Prince, although they were disappointed by the initial impression: 'The Prince, who seemed somewhat jaded, perhaps from his journey, was attired in an ordinary morning costume of black frock coat and white trousers.'[55] Once the opening ceremony was underway however, the Prince appeared in a 'magnificent robe of scarlet velvet, trimmed with ermine', which communicated his royal status in a more recognisable form.[56]

Before the Duke of Edinburgh's arrival at the Coloured Cloth Hall, a processional route was planned and the principal streets of Woodhouse Lane and the Upper Headrow were decorated by the London firm of Pigott Bros., which installed rows of Venetian masts with banners, festoons and trophies. The Venetian theme was continued by Mr E. Bachoffner along Briggate and Mr J. Whaite along Boar Lane. In addition to these 'official' decorations, homes, shops, factories and churches decorated their premises and displayed flags, flowers and Chinese lanterns, with 'the whole town wearing a gay and animated appearance'.[57] Travelling with a military escort, the Prince's carriage made its way through these

53 *Yorkshire Exhibition of Arts and Manufactures, Leeds, 1875*, p. 32.

54 'The Royal Visit to Leeds', *Leeds Mercury*, 13 May 1875, p. 7; J. Sidney Curtis, *The Story of the Marsden Mayoralty; with Sketch of the Mayor's Life. From Authentic Sources* (Leeds: T. Barmby, 1875), p. 121.

55 'The Royal Visit to Leeds', *York Herald*, 14 May 1875, pp. 6–7.

56 'The Royal Visit to Leeds', pp. 6–7.

57 'The Royal Visit to Leeds', pp. 6–7.

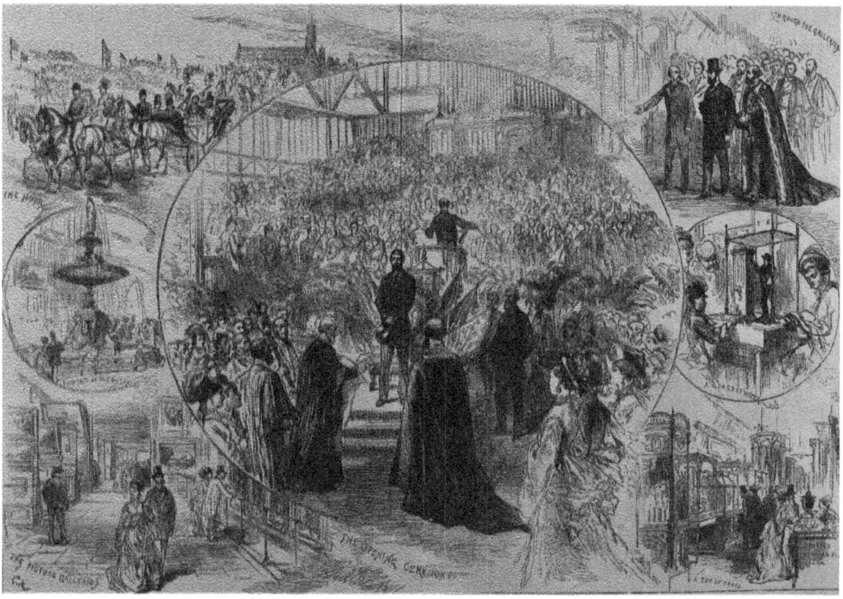

Figure 6.1 Print made by Charles Roberts, *The Duke of Edinburgh Opening the Yorkshire Exhibition of Arts and Manufactures at Leeds*, from *The Illustrated London News*, 22 May 1875, wood engraving. Collection of the author

streets to Avenue House on Woodhouse Lane, the mayor's residence just over one mile to the north. After lunch with the mayor, the party returned to the centre of the town to inaugurate the exhibition. Two thousand ticket holders had been seated in the Edinburgh Hall since just after noon, entertained by a programme of music while they waited. The audience were described as 'the *élite* of the district, attired in the height of fashion' by the *York Herald*, and the *Illustrated London News* reported that the audience was 'composed chiefly of ladies clad in the brightest colours and newest fashions'.[58] It is clear that the opening was a major social event, from the 'all ranks and classes' who lined the streets to the fashionable elite who could afford the cost of a full-price season ticket. More select still were the 900 people, 'connected more or less with the exhibition', who were invited to a grand ball in the Victoria

[58] 'The Royal Visit to Leeds', pp. 6–7; 'The Duke of Edinburgh at Leeds', *Illustrated London News*, 15 May 1875, p. 20.

Hall at Leeds Town Hall during the evening.[59] But before this finale, the Duke of Edinburgh gave an address to the audience at the Edinburgh Hall and was then led on a relatively cursory tour of the exhibition. The party left the Coloured Cloth Hall at 5 p.m. and on their return journey to Avenue House, around 30,000 children from 116 Sunday schools and societies gathered on Woodhouse Moor to sing to the Prince as he passed by.[60] The mayor paid for the production of commemorative medals for all of the children who participated, which were struck by Maher and Sons of Birmingham.[61]

The procession and its attendant events mirrored the visit made by Queen Victoria, Prince Albert and Princesses Alice and Helena to Leeds, on the occasion of the opening of the Town Hall seventeen years previously. The combination of the monarch's first visit to the town coupled with the intense interest in the new building resulted in an episode that came to represent the zenith of civic pride.[62] Gunn argued that these mid-century royal visits 'did most to stimulate large-scale civic pageantry in the industrial cities' and this was certainly true of Leeds.[63] For a moment the town was able to imagine itself as the very centre of imperial power: 'if only for a few hours, the borough became the seat of empire of the greatest monarchy of the earth'.[64] The arrival by train, the elaborate decoration of the streets, the military displays, the coalition of local aristocracy and local government, the crowds (estimated at up to 200,000 people), the processional route and the nearly 30,000 schoolchildren who were gathered on Woodhouse Moor to sing, were all emulated in 1875.[65] The events of 1858 provided both a practical model and an ideological aspiration for the organisers of the Yorkshire Exhibition of Arts and Manufactures, who sought to stimulate the collective memory of the town and wider country to set Leeds on the world stage again.

59 'Yorkshire Exhibition of Arts and Manufactures. The Duke of Edinburgh's Visit to Leeds', *Daily News*, 13 May 1875, p. 2.

60 'The Royal Visit to Leeds', *Leeds Mercury*, 13 May 1875, p. 7.

61 'The Duke of Edinburgh at Leeds', *Graphic*, 22 May 1875, p. 3.

62 'Her Majesty at Leeds', *Illustrated London News*, 11 September 1868, pp. 240–41.

63 Gunn, 'Ritual and Civic Culture in the English Industrial City, *c*. 1835–1914', p. 228.

64 'The Queen's Visit to Leeds', *Illustrated London News*, 11 September 1868, p. 244.

65 'The Queen's Visit to Leeds', p. 244.

The Manufactures Department

The Manufactures Department was located at either side of the central Edinburgh Hall, and its Section A was intended to be the first department visited, with the second part, Section D, following the two sections of the Machinery Department. The committee explained that they 'aimed at having a great variety of distinct manufactures represented, rather than a superabundance of exhibits in any particular branch' to make the display 'more interesting and valuable'.[66] Despite their aims, the committee recognised that the woollen trade continued to be important to the local area and was well represented by a variety of processes and products.[67] Allied textiles such as jute, silk and leather were also prominent in the display.[68] The *York Herald* listed some of the other manufactures, which confirmed the diversity of raw materials and commodities represented in the Manufactures Department:

> Brass finishing, nails, fire extinguishers, domestic utensils, glass, shears, knives, and cutlery; brushes, bristles, ornamental turning, decorative work, furniture, encaustic tiles; stationary calculating machine; commercial products of the vegetable kingdom used in the arts of dyeing and tanning; oils, chemicals, and drysalteries; hair and toilet requisites, &c. The food section of the department will be interesting, containing as it does cases of preserved provisions, cocoa in its various stages of manufacture from the nut, &c.; pickles, sauces, mustard, &c.[69]

Although commercial interests and imperatives were present throughout the exhibition, this department most resembled a trade show. Its primary function appeared to be to advertise local businesses and products to local people: of the eighty-eight exhibitors in Section A, fifty-seven were from Leeds. The wider Yorkshire area was well represented too, with exhibitors from Bradford, Barnsley, Castleford, Dewsbury, Halifax, Huddersfield, Sheffield and York, alongside a minority from industrial centres across Britain, including Birmingham, Glasgow, Middlesborough, Salford and Stockton-on-Tees. Many of their raw materials came from much further afield, but Britain's colonial role was largely absent from the catalogue and from wider discussion of the exhibition and its contents.

66 *Yorkshire Exhibition of Arts and Manufactures, Leeds, 1875*, p. 10.
67 *Yorkshire Exhibition of Arts and Manufactures, Leeds, 1875*, pp. 10–11.
68 *Yorkshire Exhibition of Arts and Manufactures, Leeds, 1875*, p. 11.
69 'The Royal Visit to Leeds', *York Herald*, 14 May 1875, pp. 6–7.

The eighty exhibitors of the second part of the exhibition—Section D—were subdivided into categories for leather, woollen goods and the more wide-ranging classification of 'woollens, worsteds, silks, linens, yarns, clothing, manufacturing accessories and materials, chemicals, oils, soaps, foods, etc.' An even higher proportion of the exhibitors were from Leeds in this half of the Manufactures Department, at sixty-nine, with a number having taken advantage of the opportunity to have a printed advertisement included in the official catalogue.

The Machinery Department

Once visitors had passed through Section A of the Manufactures Department they entered the Machinery Department, housed in an annex next to the Coloured Cloth Hall constructed for especially for the purpose (Figure 6.2). The timber structure was arranged over two floors, each of which was 59 metres long and 12 metres wide.[70] As had been the case in the polytechnic exhibitions of the late 1830s and 1840s, industrial machinery was presented both static and in active use. The machines shown in motion were powered by a steam engine supplied free of charge by local engineers John Fowler & Co.[71] The engine was designed for hauling and winding in underground working environments such as quarries and mines and had been repurposed here as a prime motor capable of 90–100 horsepower.[72] Across the two floors Fowler also displayed winding drums, axles, models of their famous steam ploughs and agricultural implements.[73] The firm had a presence in the Fine Art Department too, where two of their prize cups—awarded by the Viceroy of Egypt and Lorn Vernon—were displayed in a case of plate which also included Norwegian tankards and spoons and a sugar sifter.[74]

The Mayor of Leeds and principal organiser, Henry Rowland Marsden, led the seventy-four exhibitors on the ground floor with his 'New Patent Improved "Lever Stone Breaker," specially adapted for crushing hard substances, such as ores, rocks, fossils, flints &c. Fitted with self-acting screening apparatus'.[75] The latest patent for Marsden's machine was

70 'Yorkshire Exhibition of Arts and Manufactures. The Duke of Edinburgh's Visit to Leeds', *Daily News*, 13 May 1875, p. 2.

71 *Yorkshire Exhibition of Arts and Manufactures, Leeds, 1875*, p. 9.

72 'The Yorkshire Exhibition', *Leeds Mercury*, 14 May 1875, pp. 7–8 (p. 8).

73 'Yorkshire Exhibition of Arts and Manufactures at Leeds [Concluding Notice]', *English Mechanics and the World of Science*, 9 July 1875, p. 424.

74 *Yorkshire Exhibition of Arts and Manufactures, Leeds, 1875*, p. 158.

75 *Yorkshire Exhibition of Arts and Manufactures, Leeds, 1875*, p. 42.

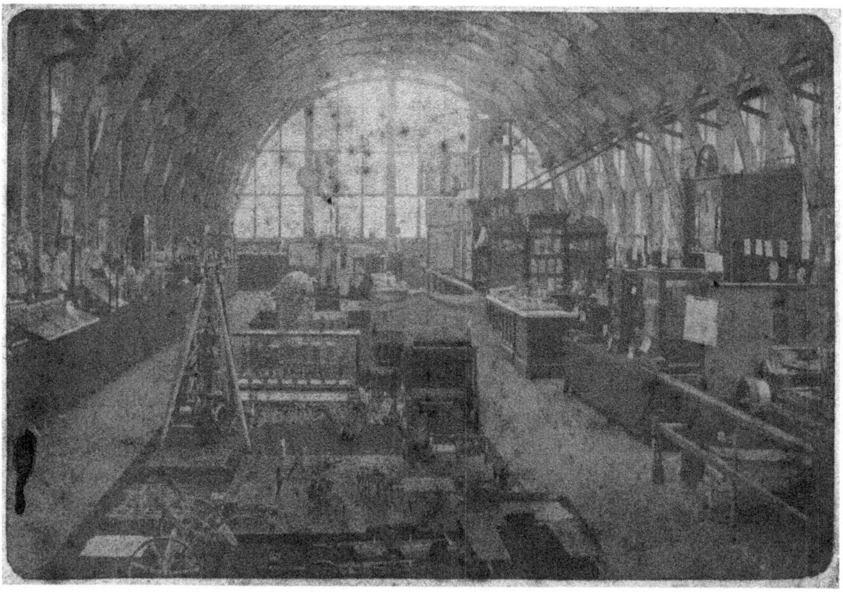

Figure 6.2 Anonymous, *Machinery Annexe C, No.1, Yorkshire Exhibition of Arts and Manufactures, Leeds 1875*, 1875, photograph. Leeds Museums and Galleries, UK/Bridgeman Images

dated 4 June 1874, so the exhibition may have been the first opportunity to demonstrate his improved design in public.[76] Marsden's biographer reported that the machine specifically caught the attention of the Duke of Edinburgh during his tour, which resulted in a demonstration of its ability to quickly crush ironstone into small fragments.[77]

Part of the upper floor was dedicated to a display of objects that had been designed or made 'by working operatives from all parts of the country', motivated by the 'substantial money prizes' offered by the Mayor of Leeds (Figure 6.3).[78] The *Leeds Mercury* assessed the display and concluded that some of the work was good enough to have been exhibited outside the context of the competition, although the review also drew attention to the 'many works to which praise cannot be

[76] No. 1937 Crushing Ores and Breaking Stones &c., *English Patents of Inventions, Specifications: 1874* (London: printed by George Edward Eyre and William Spottiswoode, 1874), pp. 1–9.

[77] Curtis, *The Story of the Marsden Mayoralty*, p. 134.

[78] Curtis, *The Story of the Marsden Mayoralty*, p. 9.

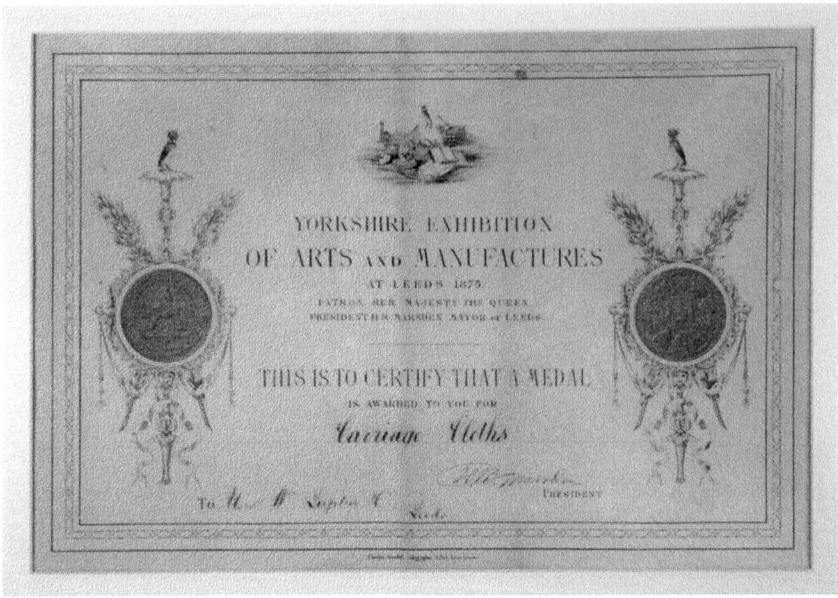

Figure 6.3 Certificate awarded to William Lupton and Company Limited, 1875. Leeds Museums and Galleries, UK/Bridgeman Images

given, unless it be on the ground that they represent a patient striving after beauty or usefulness which had failed of success through want of training'.[79] Despite the competition having been open to entries from the whole country, it was dominated by works from Leeds and the wider Yorkshire region. It also offered a context in which the work of women could be displayed. The presence of entries by two gentlemen and works of fine art disrupted the intended function of this part of the exhibition, which was to encourage invention and innovation in working operatives—to direct their limited leisure time towards a continuation of labour and use their lived experience of work as research and development for the owners of the means of production.

The periodical press beyond the local area was less enthusiastic about this section. The popular science journal *English Mechanics and the World of Science* published a review of the Machinery Department that expressed disappointment at the scale and arrangement of the display:

[79] 'The Yorkshire Exhibition', *Leeds Mercury*, 14 May 1875, p. 8.

> The machinery department can hardly be said to be worthy of the town and districts, and certainly cannot be considered as representative of local industries. In saying this we must not be understood to say that the department does not comprise much that is worthy of notice. The facts are quite otherwise. Amongst the exhibits may be found as fine examples of the highest attainment of engineering skill as in any provincial exhibition, but we cannot help thinking that a much better collection could have been got together, and one that would not give us so distinctly the impression that it is one large shop. The fault, however, is not with the exhibitors, who have done their best with the space at their command and the power supplied.[80]

The *Art Journal* was even more discontented with the Machinery Department: 'Our business, fortunately for us, is not with the painfully-confused arrangements of the machinery and industrial departments—where order it utterly absent, and where prominence is given to objects which ought to have been rejected.'[81]

The Science Department

The Science Department occupied the smallest space in the exhibition, situated on the upper floor of the Annex. It included much of what had been described in the polytechnic exhibitions of the 1830s and 1840s as 'philosophical apparatus', including telegraphic technologies, electromagnetic machines, clocks, microscopes, telescopes, barometers, surgical instruments, and photographic and chemical equipment. Like these earlier exhibitions the equipment was regularly demonstrated, although the technologies on display here were more likely to be commodities available for domestic and commercial use. Of this section, the *Leeds Mercury* concluded that it was 'an exhibition worthy of the present state of science in its practical more than its theoretical aspects. It must be admitted that they have succeeded to a greater extent than could have been expected in a provincial exhibition.'[82] The department was led by William Sykes Ward (1815–85), a solicitor and amateur scientist who was one of a small but significant number of people who had been

80 'Yorkshire Exhibition of Arts and Manufactures at Leeds [Concluding Notice]', *English Mechanics and the World of Science*, 9 July 1875, p. 424.

81 'Leeds', *Art Journal*, vol. XIV (London: Virtue & Co., 1875), pp. 221–22.

82 'The Yorkshire Exhibition', *Leeds Mercury*, 14 May 1875, p. 8.

involved in the first Leeds Public Exhibition thirty-six years earlier.[83] A contemporary commented that he 'was not possessed of great powers of organization, but was peculiarly ingenious in speculation and experiment', and further that the social position afforded to him as a solicitor 'permitted him to employ both leisure and pecuniary means in the pursuit of such arts and science as attracted his interest'.[84] An early Fellow of the Chemical Society of London and honorary secretary of the West Riding Geological and Polytechnic Society from 1855 to 1870, he delivered papers on a number of subjects during the 1840s and 1850s, including 'telegraphy, atmospheric railways, photography, and various chemical subjects'.[85] Ward pledged £20 to the guarantee fund and lent twenty-three musical instruments to the Science Department 'for the scientific investigation of musical tones and harmonics'.[86]

The Sanitary Department

The inclusion of a discrete Sanitary Department recalled the Third Leeds Polytechnic Exhibition of 1845 and its emphasis on the improvement of public health in the town. Sanitation offered an arena that sat between science and manufactures, with the invention of the latter coupled with the potential for commercial advantage that came with exhibiting new technologies in public. The sanitary situation in Leeds had improved in a number of areas since the cholera outbreaks of 1832 and 1849, with the municipalisation of the supply of piped water in 1852 and the expansion of a more effective system of sewerage and drainage.[87] In the second half of the century major infrastructure improvements to water supply were both a source of civic pride and civic rivalry, as Andrew McTominey has demonstrated with reference to Leeds and neighbouring Bradford.[88] The

83 Ward had been the honorary secretary of the Leeds Philosophical and Literary Society between 1840 and 1869. The other people involved in both the 1839 and 1875 exhibitions were James Kitson, Edward Baines and Alderman George. See 'The Royal Visit to Leeds', *Leeds Mercury*, 13 May 1875, p. 7.

84 Richard Reynolds, 'Biographical Notices', *Proceedings of the Yorkshire Geological and Polytechnic Society*, 10, 1 January 1889, 238–55 (p. 241).

85 Reynolds, 'Biographical Notices', p. 241.

86 *Yorkshire Exhibition of Arts and Manufactures, Leeds, 1875*, pp. 18, 68–69; 'The Yorkshire Exhibition', *Leeds Mercury*, 14 May 1875, p. 8.

87 David Sellers, *Hidden Beneath our Feet: The Story of Sewerage in Leeds* (Leeds: Department of Highways and Transportation, 1997), p. 13.

88 Andrew McTominey, 'Bad Neighbours? Water Supply and the Civic Rivalry

use of more hygienic water closets had increased from 1,005 in 1856 to 6,000 in 1870, although even by this date there remained around 30,000 privies in use.[89] C.J. Morgan noted that by 1870 the quantity of piped water available for Leeds households was adequate, but the quality and pressure were not yet good enough; some streets remained buried in waste and infectious diseases other than cholera continued to cause a large number of deaths.[90] The pace of positive change improved from this point onwards, however; the reduction of childhood mortality contributed to the fall in overall death rates.[91] According to B.J. Barber the number of water closets had risen to approximately 8,500 by 1875 and dramatically increased to 27,990 by 1889.[92] Exhibitions such as this were part of the apparatus of local government, with an attendant role in the improvement of public health. The promotion of sanitary technologies at the Yorkshire Exhibition of Arts and Manufactures in 1875 may have contributed to the remarkable increase in their adoption in local households. The 'soft power' of an exhibition—with its capacity to construct an argument through architecture, objects and social conformity—could perhaps effect greater and more rapid change in attitudes and behaviours.

The catalogue explained the remit of this department, which recognised both the progress that had been made over the last few decades and also the scope for further innovation and refinement:

> The development of the science of hygiene during the last quarter of a century has been accompanied ... with a considerable number of mechanical, chemical, and other appliances, for carrying into practical effect the inductions of sanitation... The various kinds of closets, disinfecting appliances, heat and ventilation producers, pipes, junctions, bricks, sewage conduits, and sewer ventilators, are doubtless capable of infinite improvement.[93]

of Leeds and Bradford, c.1850–1887', *International Journal of Regional and Local History*, 12:1 (2017), 24–41.

89 Sellers, *Hidden Beneath our Feet*, p. 13.

90 C.J. Morgan, 'Demographic Change, 1771–1911', in *A History of Modern Leeds*, ed. by Derek Fraser (Manchester: Manchester University Press, 1980), pp. 46–71 (pp. 65–66).

91 Morgan, 'Demographic Change, 1771–1911', p. 67.

92 B.J. Barber, 'Aspects of Municipal Government, 1835–1914', in *A History of Modern Leeds*, ed. by Derek Fraser (Manchester: Manchester University Press, 1980), pp. 301–26 (p. 304).

93 *Yorkshire Exhibition of Arts and Manufactures, Leeds, 1875*, p. 12.

The committee recognised that the apparatus on display in the Sanitary Department was not the most visually interesting, but had the capacity to offer the most tangible improvements to the lives of the people of Leeds.[94] In its extensive review, the *Leeds Mercury* concurred with this view: 'Although pipes, bricks, sewage, conduits, and sewer ventilators may be uninteresting objects in themselves, yet the important part they play in our sanitary system should give them sufficient interest even to the ordinary visitor.'[95]

Across fifty-one exhibitors, the majority of the objects on display were concerned with the management of smoke, ash, water, gas and waste for both domestic households and public infrastructure. The provision and quality of natural light were also concerns for the department; a significant minority of objects on display provided safety features for windows or stained glass (both real and imitation). The more decorative of these exhibits (alongside a number of photographs) were only displayed in the Sanitary Department as a result of a lack of space.[96] Although the intention had not been to display stained glass next to safety windows, the combination of these elements merged in other forms too, notably through ornamental ventilation bricks and cornicing, enamelled and glazed bricks, mouldings for picture hanging, and fireplaces, fenders and mantelpieces. These objects promised to make a safe and sanitary domestic environment something that could be purchased, at least by the middle classes.

The Fine Art Department

It was very clear that the organisers of the 1875 exhibition were not interested in positioning the event as a sequel to the National Exhibition of Works of Art of 1868. Fine art was far from the primary focus of the 1875 exhibition, and the status of this department was not elevated above the other departments in the commentary that surrounded the event. Although the Fine Art Department occupied the largest space in the North Gallery of the Coloured Cloth Hall and contained the greatest number of objects, as a category it also contained the greatest diversity of material and was expected to carry a specific narrative about human development. Despite the designation of the department, works of painting and sculpture only formed part of the display. The

94 *Yorkshire Exhibition of Arts and Manufactures, Leeds, 1875*, p. 12.

95 'The Yorkshire Exhibition', *Leeds Mercury*, 14 May 1875, p. 8.

96 'The Yorkshire Exhibition', p. 8.

official catalogue introduced the department as a space in which visitors would see 'many and varied examples of Ancient and Modern Art, important in their teaching as marking the gradual progression from the rude and savage condition of our forefathers to our present high state of civilisation'.[97] In the catalogue's introduction to the Fine Art Department, the racist implications of this ideology appeared to have been somewhat disrupted by the focus on British 'progress': from 'the ancient flints and savage implements wherewith men in those days slew their enemies, and killed wild beasts of the forest', through 'Anglo-Saxon' arms and armour to seventeenth-century furniture, ceramics and other forms of decorative arts. From there, 'Modern Art will then be seen to make its appearance, and the visitor will find himself among the furniture and fittings of his own day. Jewellery, Enamels, Ecclesiastical Art ... Continental China (ancient and modern), will be found in this collection'.[98] While the visitor had reached the present day, they had not yet come into contact with what we now understand to be works of fine art, rather a chronological (but not encyclopaedic) collection of archaeological specimens, arms and armour and decorative arts. This changed on entering the Ante-Chamber of the Fine Art Department, which contained 'the early Dutch and Italian Schools of Painting' and marked the beginning of a more recognisable display of fine art.[99]

Occasionally referred to informally as the Picture Gallery, this section continued the chronological narrative from old master works on paper to contemporary paintings. The catalogue promised that 'Art will be here represented by some of its greatest sons and greatest patrons.'[100] In accordance with the norms that unfortunately remain so entrenched in the history of Western art, it was assumed that both artists and visitors were men by default. The patrons, however, were allowed to be marginally more varied, not least because the most prestigious lender was Queen Victoria. Institutions were important lenders too, particularly the South Kensington Museum, whose collection had formed a part of the most significant temporary exhibitions in Leeds in the second half of the century. The 'nobility and gentry of the land' and 'the wealthy manufacturers of Yorkshire' were also noted as contributors; however, the sharp division between what was considered old money versus new money had been complicated by the passing of time. The longstanding wealth of those who were now second- and

[97] *Yorkshire Exhibition of Arts and Manufactures, Leeds, 1875*, pp. 5–6.

[98] *Yorkshire Exhibition of Arts and Manufactures, Leeds, 1875*, p. 6.

[99] *Yorkshire Exhibition of Arts and Manufactures, Leeds, 1875*, p. 6.

[100] *Yorkshire Exhibition of Arts and Manufactures, Leeds, 1875*, p. 6.

third-generation industrialists and merchants meant than some now had an education, estate, collection and taste almost indistinguishable from those families who could trace their wealth and titles back for centuries.

Some of the loans deviated from the narrative presented in the introduction to this section. When the visitor moved from the vestibule to the south end of Gallery F they encountered an 'ethnological collection' described as 'the Tools, Implements, and Arts of savage people, still or recently existing. Upon the positive evidence of these collections, the science of antiquities is based; as by the present the past may be judged of and realised in the remains, whether historical or pre-historic.'[101] This is the clearest expression of the racist beliefs and values embedded in the colonial project, which sought to establish an empirical basis for white supremacy through the material culture of colonised people. For a catalogue that generally only provided a number, title and contributor for each object, it is notable that the first work on display in this section was given the following extended description:

> Unique Effigy or Model of a native warrior of Raratonga [*sic*], one of the South Sea Islands. The natives of this island are of a very wild and savage disposition, as witness the murder of the Rev. Mr. Williams, of the Wesleyan Mission, and of several other massacres which have occurred there, even so late as April 1874. The whole of his "outfit" appears to be of native manufacture, and will be seen to be most fitting and suitable to meet the effects of the native weapons, viz., toothed sword, bludgeon, &c. The general net-work coat might suggest chain mail. The belt is of shark-skin; the helmet a fish-skin, and the necklace, &c., characteristic of other barbarian use of shell ornaments.[102]

More than a single object, as an effigy it operated as a proxy for a whole people and their culture. The presentation of a whole figure resulted in quite a different encounter with the visitor than if the individual elements of clothing, arms and armour and decorative art were displayed in cases as autonomous objects. The model offered context to some extent, but more than that, it imposed narrative. It materialised the 'wild and savage' warrior, denuded of a threat that was presented as contemporary. Aside from its offensive language, the passage contained basic factual errors. John Williams (1796–1839) was

[101] *Yorkshire Exhibition of Arts and Manufactures, Leeds, 1875*, p. 99.

[102] *Yorkshire Exhibition of Arts and Manufactures, Leeds, 1875*, p. 99.

a missionary in the Pacific Islands who visited Rarotonga in 1823 and 1827. He was not killed there but rather on the island of Erromango in the Vanuatu archipelago over 2,000 miles away.[103] It is possible that the writer conflated the Pacific Islands—deliberately or accidentally—into one generic place where all 'natives' were 'savage'.

The origin of this exhibit is notable: it was recorded as having been contributed by Mr R. Cundall of the Victoria Museum, Leeds. Robert Cundall (1819–96) ran his museum as a commercial venture from 1865 to 1880.[104] Located on the residential Beckett Street in the otherwise industrialised Burmantofts area east of central Leeds, the museum predominantly contained natural history specimens. Cundall advertised bird skins and mounted birds, animals and reptiles for sale at the museum.[105] The 1871 Census recorded him at age 52 living with his family at 118 Beckett Street, which was also the address of the Victoria Museum. He described his occupation as 'Museum Keeper', having been described as a 'Coal Agent & Boat Owner' ten years previously and as a greengrocer on his marriage certificate in 1849.[106] It appears that after the closure of the Victoria Museum in 1880 Cundall returned to his previous profession. The 1881 Census recorded Cundall and his wife living in the North Yorkshire coastal town of Scarborough as a 'Coal Agent'.[107] Ten years later they remained at the same address, though Cundall appears to have retired as he was described as 'living on his own means'.[108] Although he had moved away from Leeds, Cundall retained

[103] David Hilliard, 'Williams, John (1796–1839), missionary in the Pacific islands', *Oxford Dictionary of National Biography* (Oxford University Press, 2004), https://www.oxforddnb.com/view/10.1093/ref:odnb/9780198614128.001.0001/odnb-9780198614128-e-29521 (accessed 3 December 2020).

[104] Samuel Alberti, 'Field, Lab and Museum: The Practice and Place of Life Science in Yorkshire, 1870–1904' (unpublished doctoral thesis, University of Sheffield, 2000), p. 95.

[105] See advertisement in the back of J.J. Armistead, *A Short History of the Art of Pisciculture: Shewing Its Utility and Some of the Advantages which May be Derived from it If Properly Carried on* (Leeds: printed by Samuel Moxon, 1870), unpaginated.

[106] See Census Returns of England and Wales, 1871, Class: RG10; Piece: 4556; Folio: 4; Page: 2; GSU roll: 847138; Census Returns of England and Wales, 1861, Class: RG 9; Piece: 3377; Folio: 49; Page: 37; GSU roll: 543122; and Yorkshire Parish Records, Reference Number: BDP14, West Yorkshire Archive Service, Leeds.

[107] Census Returns of England and Wales, 1881, Class: RG11; Piece: 4803; Folio: 34; Page: 21; GSU roll: 1342159.

[108] Census Returns of England and Wales, 1891, Class: RG12; Piece: 3964; Folio: 120; Page: 31; GSU roll: 6099074.

links to the town through property ownership: the Electoral Register for 1887 recorded that he owned freehold houses and tenements on Beckett Street and Museum Terrace close to (and likely including) the location of the former Victoria Museum.[109] Despite his varied career, 'coal dealer' was the occupation recorded on the Beckett Street Cemetery burial on at his death aged 77 in 1896.[110]

It has not yet been possible to trace the provenance of Cundall's small ethnographic collection, which also included bows and arrows 'said to belong to the same warrior' as above and similar arms from Mexico.[111] His trade in bird, animal and reptile skins from outside Britain may have provided the opportunity to purchase or exchange other objects acquired through colonisation. What is equally difficult to determine is the ethical status of the acquisition; all we know is how it was interpreted in the catalogue. The only other mention of his loan was published in the *Leeds Mercury*, which described the warrior figure as, 'armed for battle ... his whole suit of armour corresponds well as a protection from such primitive weapons as he himself carries'.[112] What this small case study demonstrates is the importance of collectors who occupied the territory of the enthusiastic amateur without either the aristocratic title or middle-class occupation to support their activities. It shows too that the objects of Empire could be found at all levels of society and were not limited to the 'scholarly' collections of voluntary societies, universities and national museum collections.

At this stage it is important to take an extended consideration of a single collector who lent to the exhibition in order to better understand the operation and performance of taste as it relates to the public life of one of 'the wealthy manufacturers of Yorkshire' referred to in the catalogue. Joseph Walley (1813–82) had a nonconformist baptism at the Wesleyan Methodist Chapel in Holbeck.[113] He married Lydia Pickles

[109] Electoral Registers, 1840–1962, PUD:1/1, 1887, West Yorkshire Archive Service, Leeds.

[110] Beckett Street Cemetery Records 1845–1987, West Yorkshire Archive Service, Leeds.

[111] Catalogue numbers 4, 10 and 92. See *Yorkshire Exhibition of Arts and Manufactures, Leeds, 1875*, pp. 99, 101.

[112] 'The Yorkshire Exhibition', *Leeds Mercury*, 14 May 1875, pp. 6–7 (p. 6).

[113] West Yorkshire Nonconformist Records, Bradford, Calderdale, Kirklees, Leeds, and Wakefield: West Yorkshire Archive Service, Reel Number: LNC 7, Reference Number: OP 18; and The National Archives General Register Office: Registers of Births, Marriages and Deaths surrendered to the Non-parochial Registers Commissions of 1837 and 1857; Class Number: RG 4; Piece Number: 3177.

Sharp in 1850, whose father William Sharp was described as a gentleman on their marriage certificate; by contrast Walley's father William was a woolstapler.[114] The 1851 Census recorded that he lived close to the mayor at 8 Eldon Terrace with his wife Lydia and one servant and that he worked as a woolstapler as his father had.[115] His warehouse was at 7 Alfred Street off Boar Lane in the centre of Leeds.[116] He paid an annual subscription of £1 1s. for the new General Infirmary from March 1861 and continued his support when he was elected to the Weekly Board in 1877.[117] He was part of the Grand Jury for the Leeds Christmas Quarter Sessions at the Town Hall in December 1863 and the Michaelmas Quarter Sessions in October 1868.[118] In February 1868 he left the firm of Wm. Walley & Co. in Albion Street and Joseph Walley & Co. in Bond Street and the businesses were carried on by William Wilkinson Walley and Samuel Walley.[119]

Following his retirement from business, Walley increased his engagement in the public life of the town, first through visual art and voluntary societies and shortly after through local politics. In May 1869, following the National Exhibition of Works of Art in 1868, the Leeds Mechanics' Institution and Literary Society opened a picture gallery at its Institute building, to which Walley was listed as a donor.[120] The report of the council of the Leeds Philosophical and Literary Society for 1872–73 recorded Walley as a new ordinary member, paying an annual subscription of one guinea, with his wife also recorded as a 7s. 6d. subscriber.[121] He stood in the municipal election as a Liberal candidate for the Mill Hill ward in 1874 alongside Joseph Buckton. The *Leeds Times* reported that their meetings were 'most enthusiastically supported, and the general impression is that hard work, well kept up, is sure to place them in the

[114] Church of England Marriages and Banns, 1813–1935, West Yorkshire, 1850.

[115] Census Returns of England and Wales, 1851, Class: HO107; Piece: 2321; Folio: 1043; Page: 8; GSU roll: 87549-87552.

[116] UK Poll Books and Electoral Registers, 1538–1893, London Metropolitan Archives and Guildhall Library. Poll year 1860.

[117] 'General Infirmary, Leeds', *Leeds Times*, 23 March 1861, p. 4; 'Leeds General Infirmary', *Leeds Mercury*, 29 March 1877, p. 6.

[118] 'Leeds Christmas Quarter Sessions', *Leeds Mercury*, 31 December 1863, p. 4; 'Leeds Borough Sessions', *Leeds Mercury*, 27 October 1868, p. 3.

[119] Legal Notices, *Leeds Mercury*, 22 February 1868, p. 6.

[120] 'Leeds Mechanics' Institution Picture Gallery', *Leeds Mercury*, 1 June 1869, p. 5.

[121] Leeds Philosophical and Literary Society, *Fifty-Third Report of the Council of the Leeds Philosophical and Literary Society at the Close of the Session 1872–73* (Leeds: printed by McCorquodale & Co., 1873), p. 32; p. 38.

Council'.[122] Walley and Buckton won a narrow victory against the Tory candidates Thomas Walker and Thomas Turner, in an election that was described as 'a severe party struggle'. All twelve wards were contested and sixteen seats were available, of which the Liberals won eleven and the Conservatives five, shifting the composition of the council to forty-one Liberals and twenty-three Conservatives.[123] By this stage Walley was referred to as 'now out of business' and a 'gentleman' of Cumberland Road in Headingley, while his running mate was a printer from nearby Hyde Park Road. Although the council as a whole was dominated by Liberals, the Mill Hill ward was represented by four existing Conservative councillors. The Liberal *Leeds Times* reported that the 'merchants and other tradesmen having their places of business in the ward took a more lively interest in the election than has been the case for many years back, and their efforts were rewarded with complete success'.[124] Walley and Buckton were nominated by their party to stand again in the 1877 and 1880 elections, in which they retained their seats.[125] He was proposed as an Alderman in November 1880 as 'a gentleman of moderate opinions, respected and trustworthy, who had his time at his own disposal' but was not elected to this role, remaining a councillor for the rest of his life.[126] In the 1881 Census he was described as a retired wool merchant and had moved to a house in Headingley called Stony Royd with his wife, a cook and a housemaid.[127] He died aged 69 on 28 January 1882 and was interred at Lawnswood Cemetery on 2 February 1882.[128]

These biographical details are important because they establish a social, economic and political context for Walley's involvement in the Yorkshire Exhibition of Arts and Manufactures in 1875. Walley was a member of the Fine Art Department alongside Samual Walley, who was also part of the Machinery Department and chairman of the musical arrangements committee.[129] Both were also listed as guarantors at £50 each.[130] The family firm, William Walley & Co., lent 'Samples of Colonial

[122] 'Leeds Municipal Elections', *Leeds Times*, 17 October 1874, p. 5.

[123] 'Leeds', *Bradford Observer*, 3 November 1874, p. 3.

[124] 'Leeds Municipal Elections', *Leeds Times*, 7 November 1874, p. 3.

[125] 'The Leeds Municipal Elections', *Leeds Mercury*, 5 October 1877, p. 6; 'Leeds Municipal Elections', *Leeds Mercury*, 19 October 1880, p. 8.

[126] 'Election of Mayors', *Leeds Mercury*, 10 November 1880, p. 7.

[127] Census Returns of England and Wales, 1881, Class: RG11; Piece: 4538; Folio: 36; Page: 65; GSU roll: 1342092.

[128] Deaths, *Leeds Mercury*, 30 January 1882, p. 4.

[129] *Yorkshire Exhibition of Arts and Manufactures, Leeds, 1875*, pp. 5, 9, 13.

[130] *Yorkshire Exhibition of Arts and Manufactures, Leeds, 1875*, p. 17.

Wools' to the Manufactures Department.[131] Joseph Walley lent eighteen paintings to the Fine Art Department, which were listed in the catalogue as follows:

> 21. 'Landscape, with Figures and Cattle' by Berghem [Nicolaes Pieterszoon Berchem (1620–83)]
> 92. 'The Minnow Fishers' by T.M. Richardson [Thomas Miles Richardson (1784–1848)]
> 109. 'The Toilet' by E. Fichel [Benjamin Eugène Fichel (1826–95)]
> 243. 'Old, Old Story' by Midwood [William Henry Midwood (1833–88)]
> 250. 'Interior' by A. Provis [Alfred Provis (1818–90)]
> 273. 'Flower Girl' by J.H.S. Mann [Joshua Hargrave Sams Mann (c. 1826–1886)]
> 322. 'The Old Stage Wagon' by G.F. Herring Jun. [John Frederick Herring Junior (1820–1907)
> 360. 'Bank of Thames, with Cattle' by Boddington Jun. [Edwin Henry Boddington Junior (1836–1905)]
> 363a. 'The Shrimpers' by H. Shirley [Henry Shirley (active c. 1836–1870)]
> 376. 'Landscape' by John N. Rhodes [(1809–42)]
> 380. 'Punch and Judy' by Lara [Georgina Lara (1860–78)]
> 381. 'Farmyard' by H. Dell [John Henry Dell (1830–88)]
> 457. 'Poultry' by Salmon [John Cuthbert Salmon (1844–1917)]
> 790c. 'Cherubs' after Correggio
> 790d. 'Coronation of the Blessed Virgin' after Correggio
> 790g. 'Cherubs' after Correggio
> 1051. 'Shipping' by Niemann [Edmund John Niemann (1813–76)]
> 1108. 'Venetian Scene' by an unknown artist

This combination of contemporary British genre and landscape paintings, a seventeenth-century Dutch landscape and copies after the sixteenth-century Italian painter Antonio da Correggio reflected wider practices of collecting and display during this period. Walley was part of a second generation of middle-class collectors for whom the acquisition of contemporary British art was a priority.[132] For Giles Waterfield, they 'enunciated a new approach to collecting, celebrating narrative, geographical documentation, and civic identity'.[133] Narrative

131 *Yorkshire Exhibition of Arts and Manufactures, Leeds, 1875*, p. 81.

132 Amy Woodson-Boulton, *Transformative Beauty: Art Museums in Industrial Britain* (Palo Alto, CA: Stanford University Press, 2012), p. 10.

133 Giles Waterfield, *The People's Galleries* (New Haven, CT: Yale University

and landscape were certainly well represented in the selection that Walley lent to the exhibition, although local artists and scenes were not the most prominent in the selection.

It is not clear when Walley began to collect paintings; it is noteworthy that he was not listed as a lender to the National Exhibition of Works of Art in 1868—despite his longstanding support for the Infirmary—which suggests that he may have begun after his retirement earlier in the same year. His election to the Town Council in 1874 was arguably a significant motivation for his involvement as a guarantor, member of the committee and lender, led as it was by the Mayor of Leeds. The exhibition was a vehicle for Walley and his peers to express their participation in the public life of the town and to demonstrate that commercial men had a stake in the social role of the fine arts.

The confluence of these activities aligns with the historian Richard Trainor's reassessment of urban governance in the second half of the century. Trainor questioned the prevailing narrative that the social status of councillors declined during this period, emphasising instead the 'adaptation and diversification' of local government in industrial urban centres.[134] Despite the increasing migration of the upper middle classes away from the urban centres, Trainor concluded: 'there were still many individuals who combined high socio-economic rank with a considerable public presence and were influentially active in a number of major public spheres'.[135] The involvement of people such as Walley in the Yorkshire Exhibition of Arts and Manufactures demonstrates that this was true of Leeds in the last quarter of the nineteenth century.

From the consideration of a single collector, we narrow our focus further to a single painting displayed at the exhibition, to ask questions about attribution and the transfer of works of art from private ownership through to public display and finally into municipal collections. Like Walley, Lancelot Wedderburn (1819–97)—Weatherburn in his earlier years—was also a member of the committee of the Fine Art Department, a guarantor and a prolific lender of works of fine and decorative arts to the exhibition.[136] Born in Thirsk in North Yorkshire,

Press, 2016), p. 201.

134 Richard H. Trainor, 'The "Decline" of British Urban Governance since 1850: A Reassessment', in *Urban Governance: Britain and Beyond since 1750*, ed. by Robert J. Morris and Richard H. Trainor (Aldershot: Ashgate, 2000), pp. 28–46 (p. 32).

135 Trainor, 'The "Decline" of British Urban Governance since 1850', p. 34.

136 *Yorkshire Exhibition of Arts and Manufactures, Leeds, 1875*, pp. 5, 17. Wedderburn contributed £50 to the guarantee fund.

Wedderburn described himself as a tobacco manufacturer, stockbroker, accountant and, later, a gentleman.[137] He lived for most of his adult life at Hope House in Moor Allerton, then an agricultural hamlet around three miles north of central Leeds.[138] Wedderburn appears not to have entered into local politics or public life to the same extent as Walley, although he was recorded as part of the committee to re-elect the Conservative councillor for the Headingley Ward—William Lawies Jackson (1840–1917)—in the 1872 municipal election, which placed him on the opposite side of the political divide to Walley.[139] Wedderburn had a longer history of lending works of art to temporary public exhibitions in Leeds, having contributed pictures to the Leeds Philosophical Hall on the occasion of the inauguration of its extension and remodelling in 1862; the Chapel Allerton and Potternewton Mechanics' Institute in 1863; the National Exhibition of Works of Art in 1868; and the Picture Gallery at the Mechanics' Institution in 1869.[140]

[137] Wedderburn's father, also Lancelot (1789–1878), was baptised in Ponteland in Northumberland, married in Thirsk in North Yorkshire and died in Huddersfield in West Yorkshire, where he worked as a tobacco manufacturer, having previously been 'a sea captain in the transport service'. See Alexander Wedderburn, *The Wedderburn Book*, vol. 1 (Printed for private circulation, 1898), p. 504. At the time of his marriage to Lucinda Turlay on 27 March 1839, Lancelot Weatherburn Junior [*sic*] recorded his profession as 'Tobacco Manufacturer', in Huddersfield and his father's profession as 'Linen Draper'. See Church of England Marriages and Banns, 1813–1935, West Yorkshire Archive Service, Leeds, Yorkshire Parish Records; Reference Number: RDP68/5/5. His partnership with Thomas Rhodes as stockbrokers in Huddersfield was dissolved on 28 March 1845. See 'Partnerships Dissolved', *Perry's Bankrupt Gazette*, 5 April 1845, p. 7. A similar partnership with William Jennings in Leeds was dissolved in 1848. See 'Partnerships Dissolved', *Leeds Times*, 12 February 1848, p. 2. He was still recorded as a 'Stock Broker', in the 1851 census. See Census Returns of England and Wales, 1851, Class: HO107; Piece: 2316; Folio: 105; Page: 24; GSU roll: 87536-87537. He was recorded as a 'Gentleman', in the 1861 census. See Census Returns of England and Wales, 1861, Class: RG 9; Piece: 3357; Folio: 26; Page: 46; GSU roll: 543119. By the time of the 1871 census he had reverted to the original spelling of his surname, remarried after the death of his first wife and his occupation was recorded as 'Accountant'. See Census Returns of England and Wales, 1861, Class: RG10; Piece: 4574; Folio: 21; Page: 36; GSU roll: 847251. In his final appearance in the census in 1891 he was recorded as 'Living on his own means'. See Census Returns of England and Wales, 1891; Class: RG12; Piece: 3713; Folio: 15; Page: 23; GSU roll: 6098823.

[138] William White, *White's Directory of Leeds* (Leeds, 1872), p. 335.

[139] 'Headingley Ward', *Leeds Mercury*, 26 October 1872, p. 12.

[140] 'Leeds Philosophical and Literary Society', *Leeds Mercury*, 17 December 1862,

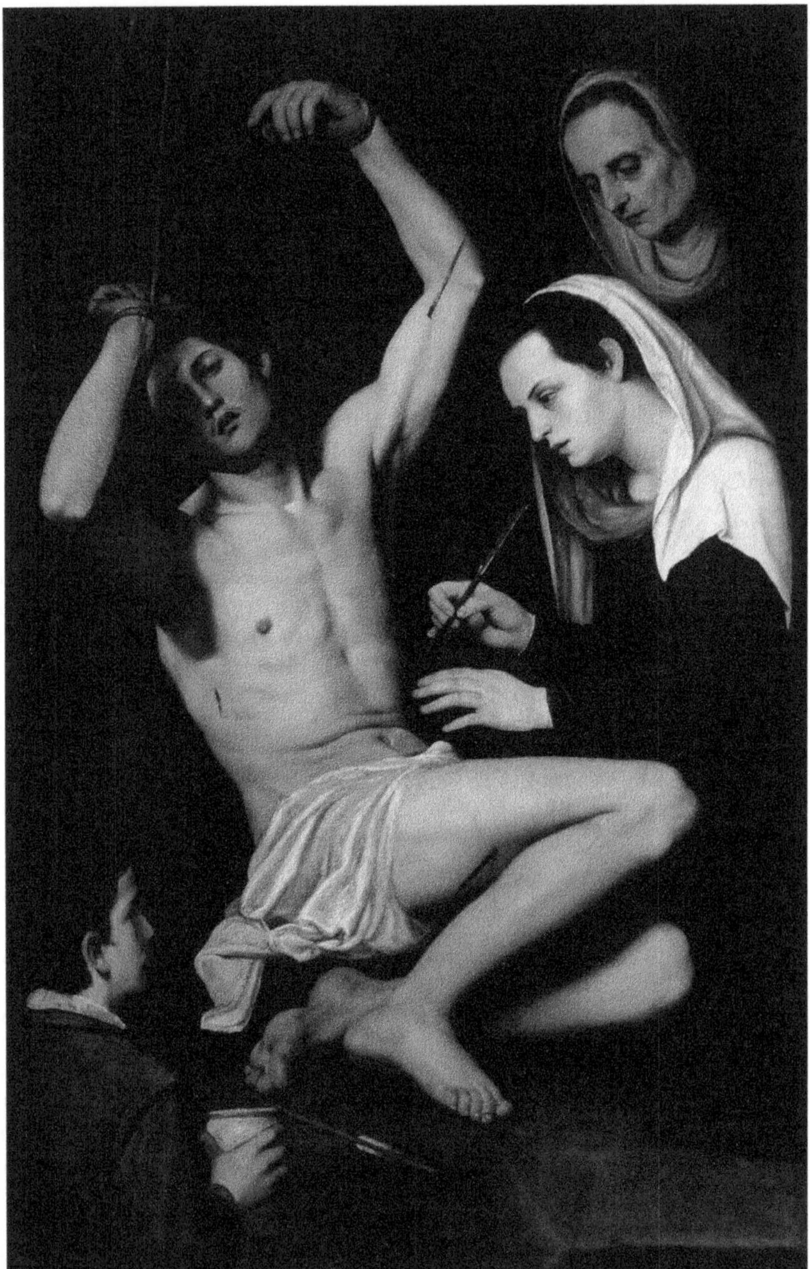

Figure 6.4 Italian School, *Saint Sebastian Tended by the Holy Women*, 17th century, oil on canvas, 168.9 × 108 cm. Leeds Museums and Galleries, UK/Bridgeman Images

Object number 8a in Gallery G of the Fine Art Department was listed in the catalogue as 'The Martyrdom of S. Sebastian, taken from a Convent in Naples during the wars of the Old Napoleon' by Murillo, loaned by L. Wedderburn (Figure 6.4).[141] This was not the first time, however, that the painting had been on public display in Leeds, having first been shown just over half a century earlier in 1824 as part of an exhibition staged by the Northern Society for the Encouragement of the Fine Arts.[142] The work presumably entered the country after the end of the Napoleonic Wars in 1815, although its provenance before 1824 is unrecorded. If it was removed from a convent in Naples during the Napoleonic Wars, the painting could be considered to have been looted twice over: by the French and then by the British after their victory at Waterloo. That the painting was the cultural property of Naples—and specifically a religious order—appears not to have been recognised, as the theft was legitimised as the spoils of war. It was intended to be a symbolic injury to the French rather than the original Neapolitan owners of the picture, hence the references in the catalogue entries to its place of origin.

After 1824 the painting was loaned to the Leeds Public Exhibitions of 1839 and 1845 by its then owner, John Bischoff (1775–1845), a woollen manufacturer and founder member of the Leeds Philosophical Society.[143] At the 1839 exhibition it was hung in the Picture Gallery, with the catalogue recording it as having come 'from a Convent in Naples' and ambitiously attributing it to Caravaggio.[144] In the 1845 exhibition it was displayed in the Saloon as a work by 'M.A. Caravaggio'.[145] At some point after Bischoff's death in 1845 Wedderburn acquired the painting, which he lent to the Leeds

p. 3; 'Chapel Allerton and Potternewton Mechanics' Institute.—Interesting Exhibition', *Leeds Intelligencer*, 4 July 1863, p. 5; 'Leeds Mechanics' Institution Picture Gallery', *Leeds Mercury*, 1 June 1869, p. 5.

141 *Yorkshire Exhibition of Arts and Manufactures, Leeds, 1875*, p. 202.

142 *Catalogue of Pictures, by the Ancient Masters in the Gallery of the Northern Society for the Encouragement of the Fine Arts* (Leeds: William Baines, 1824).

143 Edwin Kitson Clark, *The History of 100 Years of Life of the Leeds Philosophical and Literary Society* (Leeds: Jowett and Sowry, 1924), p. 12.

144 No. 156 in *Catalogue of the Exhibition of Paintings, Curiosities, Models, Apparatus, and Specimens of Nature and Art* (Leeds: Edward Baines and Sons, 1839), p. 10.

145 No. 14 in Edward Hailstone, ed., *Catalogue of the Leeds Third Polytechnic Exhibition, for Providing Public Walks and Baths, containing a magnificent collection of Sculpture, Cartoons, Paintings, Antiquities, Curiosities, Philosophical Apparatus, Mechanical Models, &c. with the names of the contributors, opened June 9th, 1845, at the Music-Hall* (Leeds: printed by Bolland & Kemplay, 1845), p. 40.

Philosophical and Literary Society for display in the Philosophical Hall from 16 December 1862 to 31 January 1863.[146]

Before his death in 1897 Wedderburn left the painting to Leeds Corporation.[147] The picture was transferred to the then relatively new Leeds Art Gallery, although it appears not to have been documented as part of the collection until 1933 and the arrival of John Rothenstein (1901–92) as the new director. The *Manchester Guardian* reported that Rothenstein had discovered the painting in the cellar of the gallery, and that it 'had been lying there unnoticed for over fifty years'.[148] Although it had been acquired as a work by the Spanish baroque artist Bartolomé Esteban Murillo (1617–82), Rothenstein consulted a number of unnamed experts and concluded that it was by the Spanish artist Jusepe de Ribera (1591–1652). Over thirty years after its rediscovery Rothenstein recalled the incident in his autobiography:

> Beneath the galleries were two large cellars in which paintings and sculpture not required for exhibition were stored. Here, mostly undisturbed for nearly thirty years, was a miscellaneous accumulation of paintings, packing cases and plaster casts, some of these last as much as eight feet high and thick-blanketed in black Leeds soot. By spring, 1933, both were clean and well lit, and the grubby hoard that had lain there had been sorted and arranged. This Augean stable yielded a number of works of considerable interest, notably a big Riberesque 'Martyrdom of St. Sebastian'. When the picture came to light I first took it for an empty frame, so tightly packed with soot was the space between the picture surface and glass. Investigation established that it had been exhibited in 1875 at the Yorkshire Exhibition of Arts and Manufactures and afterwards presented to the Corporation by Mr. L. Wedderburn, but it had never been exhibited and there was no mention of it in any Gallery record or publication. This discovery attracted wide attention and it has been on exhibition

146 Kitson Clark, *100 Years of Life of the Leeds Philosophical and Literary Society*, pp. 74–75.

147 He had effects of £1554 12s. at the time of his death. Principal Probate Registry, Calendar of the Grants of Probate and Letters of Administration made in the Probate Registries of the High Court of Justice in England, p. 88.

148 'Leeds Art Gallery Discovery: Work of a Spanish Master?', *Manchester Guardian*, 28 January 1933, p. 7.

> ever since. Although previously attributed to G.B. Caracciolo, it is now attributed to Ribera himself.[149]

Even after its rediscovery and renewed scholarly attention the attribution remains unstable. Now referred to by the title *St Sebastian Tended by the Holy Women*, the painting is more generally credited to the Italian School.[150] It is possible that the shifting consensus on its authorship represented an improvement of art historical understanding as the century progressed, with a tendency to move away from attaching canonical artists such as Caravaggio to art objects without sufficient evidence. As Waterfield noted in his history of municipal art galleries in Britain, the quantity and quality of history of art publications increased during the second half of the nineteenth century as the discipline developed across Europe.[151]

Farnley Hall

The Yorkshire Exhibition of Arts and Manufactures made deliberate reference back to the exhibitions of 1839 to 1845 through its polytechnic system of classification and financial support for the Leeds Mechanics' Institution. Beyond these organising principles, the 1875 exhibition recalled its precursors through familial connections. Francis Hawksworth Fawkes (1797–1871) played a prominent role in these early exhibitions through the loan of watercolours by Turner and his collection of Fairfaxiana. Ayscough Fawkes (1831–99) inherited Farnley Hall and its contents in 1871 and instead of following the model of lending pictures and objects, he instead invited up to fifty season ticket holders to visit Farnley Hall each Friday afternoon between 2.30 p.m. and 5 p.m. until the end of August 1875. The opportunity was not entirely limited to season ticket holders as single tickets costing 2s. each were available to purchase through the curator Mr Ponsonby. The opportunity was advertised as follows in the official catalogue:

> Connoisseurs will be especially interested in visiting this place, on account of its being the spot where the great Turner spent

[149] John Rothenstein, *Summer's Lease: Autobiography, 1901–1938* (New York: Holt, Rinehart and Winston, 1965), pp. 195–96.

[150] Collection of Leeds Museums and Galleries, Accession Number LEEAG. PA.1907.0001.0001.

[151] Waterfield, *The People's Galleries*, p. 122.

> so much of his time, and that during the best period of his life. In this fine old house will be found in the Saloon forty-seven Water Colours by G.J.M.W. Turner [*sic*], and the Drawing-Room and Library contain six oil pictures by the same painter, and also by the following masters:—Vandyke [*sic*], Rubens, Snyders, Correggio, Guido, Sir J. Reynolds, and Romney. In the Entrance-Hall—Snyders, also Cromwellian relics, &c.[152]

It is not clear why Ayscough Fawkes decided not to lend works from the collection and instead invited ticket holders to visit his home. Certainly it reduced the inherent risks associated with handling, packing, transport and public display. These visits also referred back to the practice more readily associated with the eighteenth and early nineteenth centuries, of owners of country houses opening their doors to select visitors of their own, or comparable, class and education.[153] This tradition was reflected in the call to 'connoisseurs' to visit Farnley Hall; this was not an open invitation to the working classes whose patronage was sought for the main exhibition at the Coloured Cloth Hall.

Conclusion

The Yorkshire Exhibition of Arts and Manufactures closed on Saturday 30 October 1875. The *Bradford Observer* reported that the last day was particularly well attended, 'and as evening came on the throng became so dense that an examination of the objects of interest was entirely out of the question'.[154] In his capacity as president of the exhibition committee, the mayor addressed the crowd on the occasion of the closure. Speaking after the conclusion of a programme of music, he declared: 'a glorious battle had been fought, and the victory won (applause). That which they began eighteen months ago with fear and trembling had now been realised (hear, hear)... It was a grand and glorious success, and he had no doubt a large amount of this success was due to the exhibition being held in that part of the town.'[155] He confidently announced that the Leeds Mechanics' Institution and Literary Society had been freed from debt, with 430,618 visitors having

152 *Yorkshire Exhibition of Arts and Manufactures, Leeds, 1875*, p. 201.
153 Waterfield, *The People's Galleries*, p. 15.
154 'Close of the Yorkshire Exhibition', *Bradford Observer*, 1 November 1875, p. 3.
155 'Close of the Yorkshire Exhibition', p. 3.

paid for entry, with a further 211,516 visits by season ticket holders. As a result the mayor estimated the total gross takings to have been £34,000, predicted to be enough to pay off the debt and leave approximately £4,000 over.[156] The journal *British Architect* provided further figures that anticipated expenses of approximately £20,000 and a profit of £12,000 before the debt was paid. The article reported that the surplus above the amount required to clear the debt 'will probably be appropriated to the formation of a permanent Fine Art Gallery'.[157] The *Academy* also reported on the possibility of devoting the surplus to this purpose, which it thought 'would be a great boon to the inhabitants of Leeds'.[158]

Just over a year later in November 1876 the official financial results were issued. Although the total receipts came to an impressive £33,791. 17s. 4d., the expenses had been greater than predicted at £26,782. 5s. 6d. The remaining profit of £7,009 11s. 10d. was certainly a respectable return, but it did not quite meet the £9,500 required to completely clear the debt against the Leeds Institute building and did not leave a surplus for other purposes.[159] Although the Yorkshire Exhibition of Arts and Manufactures was intended to benefit the Leeds Mechanics' Institution and Literary Society by liquidating its debt, the exhibition had an afterlife in other ways. The Leeds Philosophical and Literary Society purchased examples of Cypriot pottery loaned by Thomas Backhouse Sandwith (1831–1900), who had built a substantial collection of antiquities during his tenure as Vice-Consul in Cyprus from 1865 to 1870.[160] The Society took the opportunity to purchase display furniture from the exhibition too, in the form of three large ebonised cases intended for use in the Mammalian Room of the museum in

[156] 'Close of the Yorkshire Exhibition', p. 3.

[157] 'Yorkshire Exhibition of Arts and Manufactures', *British Architect*, 5 November 1875, p. 260.

[158] 'Notes and News', *Academy*, 6 November 1875, p. 489.

[159] 'The amount taken by season tickets was £4375. The admission on halfcrown days realised £1416; on shilling days, £19,741; five shilling tickets, £45; admission on special days, £254; admission at 6d. after six in the evening, £11.10s. Other receipts bring the total up to £33,791. 17s. 4d. The payments amount to £26,782. 5s. 6d., thus leaving a balance to the good of £7009 11s. 10d.' See 'Result of the Recent Exhibition of Arts and Manufactures in Leeds', *Sheffield Independent*, 30 November 1876, p. 6.

[160] Leeds Philosophical and Literary Society, *Fifty-Sixth Report of the Council of the Leeds Philosophical and Literary Society, at the Close of the Session 1875–76. Read at the Annual Meeting, May 2nd, 1876* (Leeds: Printed at the Office of the Yorkshire Post, 1876), p. 9.

the Philosophical Hall.[161] The museum saw its largest ever number of visitors during the year 1875–76, at 51,112 admissions, 'no doubt partly due to the strangers brought to the town' by the exhibition that took place only a couple of minutes' walk away.[162]

Recalling the polytechnic public exhibitions of the previous generation, the Yorkshire Exhibition of Arts and Manufactures revivified the form, presented on a much larger scale. It was the first exhibition in the town to display explicit municipal governance through the Mayor of Leeds, which established a precedent for the more substantial involvement of the Town Council in the cultural sphere. Although the Leeds Mechanics' Institution and Literary Society maintained an active involvement in the organisation of the exhibition—particularly as it stood to benefit from its profits—it marked an important transition that would result in a municipal art gallery for Leeds in the following decade. The Coloured Cloth Hall provided a central space, close to connections with the national railway network, that could accommodate large numbers of objects and people. Despite its outmoded architecture evoking an earlier period of the textiles trade, the construction of the Edinburgh Hall in its substantial courtyard ensured that the interior communicated the contemporary moment. By substituting what would usually have been a bespoke iron structure for standardised wood, the expenses that had overwhelmed the National Exhibition of Arts in 1868 were avoided. Having secured the patronage of Queen Victoria and a royal visit from the Duke of Edinburgh, the exhibition and its opening ceremony took its cues from the Queen's visit to Leeds in 1858. The exhibition reprised the sense of civic pride that had been a distinctive characteristic of the 1850s, while the performance of power reaffirmed the authority of both the sovereign and the mechanisms of local government.

With its return to the presentation of manufactures, machinery, science and sanitary ware, the exhibition engaged with the industrial and commercial interests of the town and the improvements in public health related to new water and drainage infrastructure. Unlike the polytechnic exhibitions of the previous generation, the different departments of the Yorkshire Exhibition of Arts and Manufactures were more discrete, at least partially as a result of increased space. The Fine Art Department represented an unusually broad set of categories, in

[161] *Fifty-Sixth Report of the Council of the Leeds Philosophical and Literary Society*, p. 1e Leedsico.

[162] *Fifty-Sixth Report of the Council of the Leeds Philosophical and Literary Society*, p. 10.

order to construct a particular narrative of human progress predicated on the perceived contrast between objects designated as archaeological or ethnographic and works of fine and decorative art. Civic leaders were important lenders to this exhibition, and the involvement of local collectors—who had made their money in the town before entering into politics—set the stage for the final endeavour to establish a permanent art gallery for Leeds.

7

Towards an Art Gallery for Leeds, 1858–88

The call for a permanent art gallery for Leeds had been a consistent feature of the discussion around the temporary exhibitions staged in the town throughout the nineteenth century. The impetus came from two directions: the belief in the capacity of art to educate and ennoble and the concern that Leeds had failed to compete with its regional neighbours in the establishment of such an institution. These twin motivations—a positive desire and a negative fear—are a feature of almost all the local published debate on the subject, which tended to emerge with greater strength and urgency just after a temporary exhibition had taken place. The pitch of the debate was influenced by the perceived success or failure of the exhibition in question. If the exhibition had been perceived a success, the newspapers would urge the town to seize the momentum of the event and take encouragement from the support of the public. Conversely, if the exhibition had been judged a failure, the public discourse blamed the lack of interest on the absence of commitment to the visual arts in the form of a permanent gallery.

A purpose-built art gallery was so firmly positioned as a social, political, economic and cultural panacea for Leeds that it is easy to be carried along by the narrative. The relatively belated establishment of Leeds Art Gallery appeared to be such a sore point among a vocal and visible group of people that a sense of cultural lack became deeply embedded in the town's collective identity. The persistence of this story contributed to the erasure of the significance of the temporary exhibitions held before the opening of the gallery in 1888, which this book has sought to redress. For this reason it is important to note that although this chapter comes towards the end of this study, it is not intended to suggest that the formation of Leeds Art Gallery was the inevitable conclusion of the activities of the previous fifty years. Certainly it was informed by these exhibitions, but the path towards the art gallery was not linear, and it should not be thought of as the apex of cultural achievement, significant though it was and continues to be.

In contrast to the material and symbolic solidity of a permanent art gallery (the term 'permanent' was used extensively in the periodical press at the time, although we might question the ultimate permanence of any given structure or institution), temporary exhibitions are too easily interpreted as provisional and ephemeral, both then and now. As the construction of a purpose-built art gallery was a stated aim of these exhibitions—although often one of many—it is relevant to consider the ways in which they contributed to this outcome, while preserving their particular significance. Public calls for an art gallery emerged around the same time as public exhibitions in Leeds, during the opening years of the nineteenth century, but instead of tracing each instance, this chapter begins with the first concrete attempt to form a civic collection. Appended to the opportunity presented by the new Leeds Town Hall in 1858, the opening section of this chapter considers the manoeuvres that resulted in the donation of the first paintings and sculptures explicitly to the town and its people. Moving ahead by a decade, we go on to analyse the effect of the National Exhibition of Works of Art in 1868, an object of which was to stimulate the foundation of a permanent art gallery in Leeds. Into the 1870s we witness the activities of what would become the Yorkshire Fine Art Society, whose exhibitions in the Athenæum Building both recalled the miscalculations of earlier exhibiting groups and prepared the ground for those who would achieve the aim of constructing a purpose-built art gallery only a few years later.

The establishment of a civic collection at Leeds Town Hall

In the weeks leading to the opening of the new Leeds Town Hall on 7 September 1858, a group of men connected to the political, commercial and cultural life of the borough began to discuss the possibility of appropriating space in the building to be used as a public art gallery (Figure 7.1). Meeting first at the bank of Beckett & Co. on 5 August and then again at the Philosophical Hall on 9 August, the mayor, Peter Fairbairn, led a discussion of the ways in which the town could be improved through the acquisition and display of art.[1] It was suggested that the means by which these improvements could be achieved was through an annual subscription to fund the purchase of paintings and

1 'Proposed Fine Art Gallery in the Leeds Town Hall', *Leeds Mercury*, 10 August 1858, p. 3.

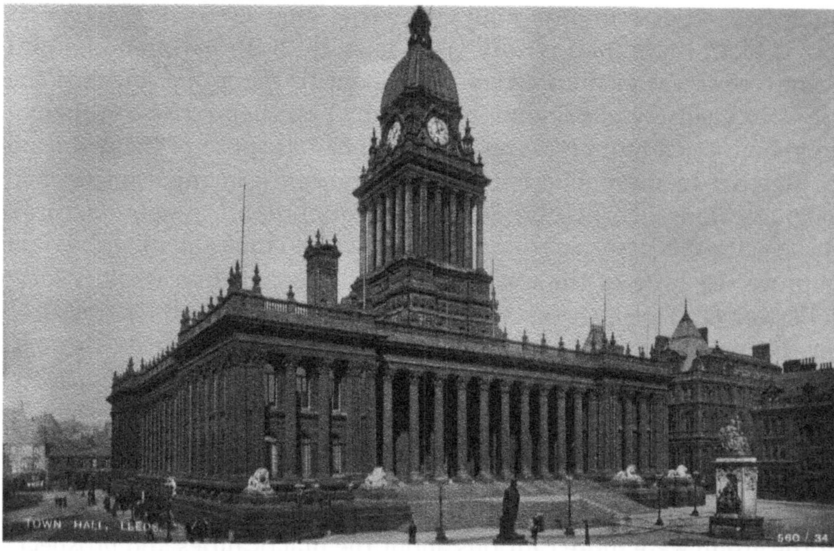

Figure 7.1 William Ritchie and Sons Limited, *Leeds Town Hall*, c. 1913, photographic postcard. Collection of the author

sculpture to be shown in the Town Hall. Donations were also strongly encouraged, with the incentive that the names of donors would be attached to the works of art. Fairbairn argued that 'the love for the beautiful should be cultivated if it were to have any practical results, and this could only be done by affording the masses of the people an opportunity for inspecting works of art of a superior character'.[2] He thought that the Town Hall could operate as an informal school of art, to teach form and colour to artisans towards the improvement of their work and, by extension, the improvement of the economic status of the town. Fairbairn expressed the logic of the design reform movement, familiar in Leeds since the first drive to establish a School of Design in the early 1840s and given additional impetus by the Great Exhibition of the Works of Industry of All Nations in 1851 and the more recent completion of the transfer of the former Museum of Manufactures to South Kensington in 1857.

Familiar too was the argument that provision for art education was superior in Europe, which resulted in improved manufactures and cultivated consumers. This position had been codified by the Select Committee on Arts and Manufactures led by William Ewart, which

[2] 'Proposed Fine Art Gallery in the Leeds Town Hall', p. 3.

published reports in 1835 and 1836.[3] The language of these reports was evident more than twenty years later. Both Fairbairn and Edward Baines, who was also an active participant in the group, invoked the idea that Leeds—and Britain more broadly—was behind its continental neighbours in a number of related ways. Baines spoke of the inclusion of high art in the public buildings of France, Germany, Holland and Belgium, alongside their profitable application of art to manufactures. He specifically asserted that a permanent exhibition of art would improve the taste of the working classes, and called on wealthy local collectors (and even Queen Victoria herself) to transfer their treasures for this purpose.[4] Baines appealed to their wish for posterity through the permanent preservation of their name and praised Fairbairn for having set an example through his donation of a marble statue of Queen Victoria by the sculptor Matthew Noble, which remains on display at Leeds Town Hall (Figure 7.2).

Despite the forceful advocacy of Baines and Fairbairn, other members of the group were more cautious about the practicalities and prospects of the idea. The question of public access was raised, dismissed by Fairbairn as 'only a matter of arrangement', and raised again, which resulted in the acceptance that a permanent arrangement could not be guaranteed by a Town Council that was subject to regular change of personnel and political priorities.[5] The prominent physician John Deakin Heaton presented a compromise position by suggesting that the Town Hall be used as a temporary art gallery through which to form the nucleus of a permanent collection. Heaton expressed discomfort at the prospect of such a collection being brought under the custodianship of the Town Council, suggesting instead that donors would prefer the control of a board of trustees.[6] The historian Simon Morgan articulated Heaton's investment in and active role in the construction of civic pride—both in public and private, as an individual and a member of the educated elite—including his advocacy of the new Town Hall.[7] Morgan reveals Heaton to have influenced the development of the town through

3 William Ewart, *Report from Select Committee on Arts and Manufactures: Together with the Minutes of Evidence, and Appendix* (London: House of Commons Papers, 1835), and William Ewart, *Report from the Select Committee on Arts and their Connexion with Manufactures; with the Minutes of Evidence, Appendix and Index* (London: House of Commons Papers, 1836).

4 'Proposed Fine Art Gallery in the Leeds Town Hall', p. 3.

5 'Proposed Fine Art Gallery in the Leeds Town Hall', p. 3.

6 'Proposed Fine Art Gallery in the Leeds Town Hall', p. 3.

7 Simon Morgan, 'John Deakin Heaton and the "elusive civic pride of the Victorian middle class"', *Urban History*, 45:4 (2018), 595–615 (p. 595).

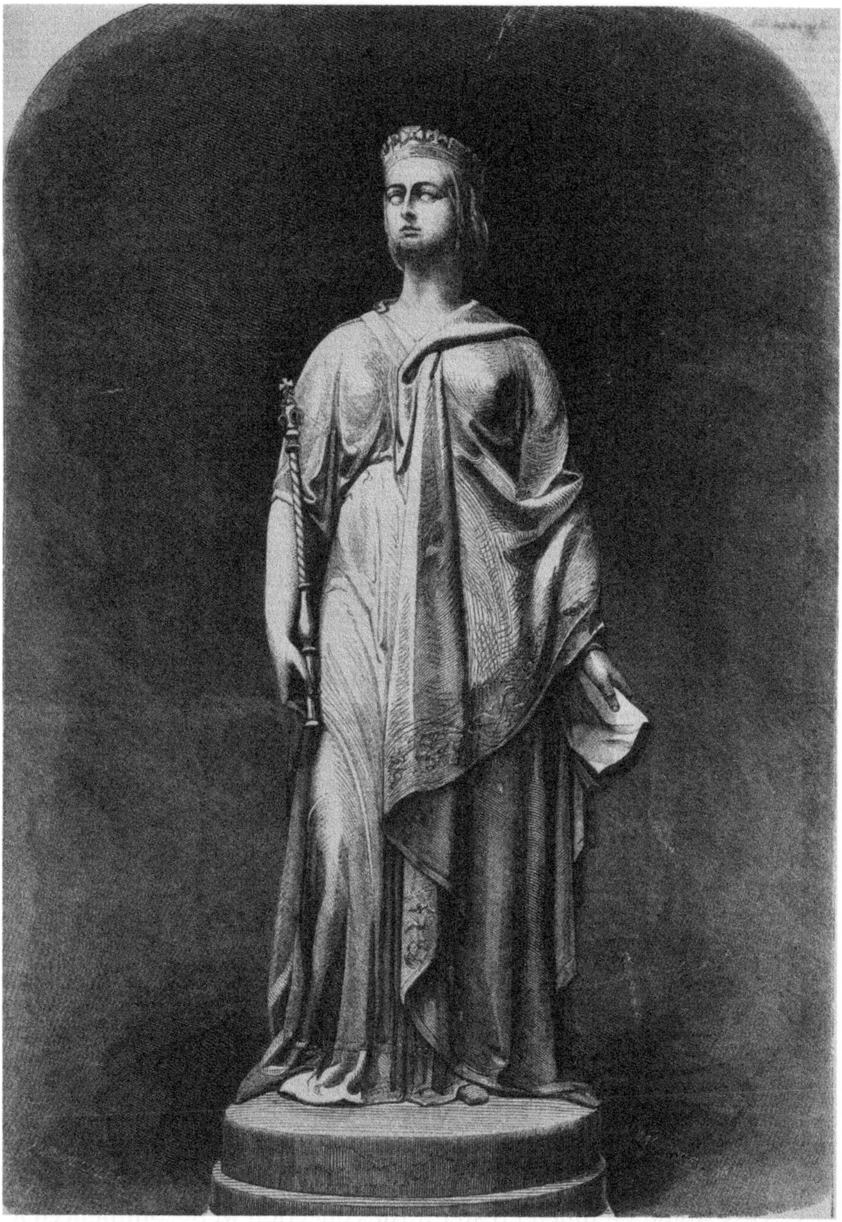

Figure 7.2 Print made by John Walmsley after Matthew Noble, block cut by Harden Sidney Melville, *Statue of Queen Victoria in the New Towhhall [sic] Leeds*, from *The Illustrated London News*, 11 September 1858, wood engraving. Collection of the author

his 'associational connections' rather than direct involvement in local government.[8] Many of the men who supported the scheme alongside Heaton had long been leading and even foundational members of local voluntary societies, including the Leeds Philosophical and Literary Society whose building they occupied for this meeting. They were often motivated by a sense of social obligation conferred by their economic and intellectual status, combined with the kind of self-interest that could be presented as philanthropy or public spirit.

In order to advance the argument, Baines claimed that he knew several men ready to make significant donations of works of art. He supported the idea of the Town Hall as a location for these donations, noting the high cost of the building, its status as a public space belonging to the ratepayers and the unlikelihood of the town being able to construct a new purpose-built art gallery. Baines thought that rooms in the Town Hall could be open to the public on selected days, with the possibility of the hall keeper showing visitors around. Such limited, supervised provision would have diverged from the intended accessibility of the public art gallery on the model of the South Kensington Museum. That some members of the group were willing to pursue this idea revealed a fundamental division that would colour the debate for the next thirty years: those who wanted to do it now versus those who wanted to do it well. Of course, the argument was not quite so binary, with the compromise position often suggested to do it now and then do it well later. At this point, however, the group could only agree to take their provisional propositions to the Town Council, and six men were selected to begin these discussions.[9]

In his capacity as mayor, Fairbairn brought the subject before the Town Council within a few days. He reinforced the favourable circumstances connected with the opening of the new Town Hall and expressed his confidence that those who had made their money in Leeds would contribute works of art to the cause. The twin benefits of education and recreation were promoted, alongside the customary belief that object lessons in form and colour, delivered to working people through exposure to art, could be translated into profit for their employers.[10] Although he recognised that the formation of a strong collection could take decades, Fairbairn stressed the urgency of the task: 'the chances of success would be much greater if the collection were commenced at

[8] Morgan, 'John Deakin Heaton', p. 597, p. 604, p. 613.
[9] 'Proposed Fine Art Gallery in the Leeds Town Hall', p. 3.
[10] 'Proposed Fine Art Gallery for Leeds', *Leeds Times*, 14 August 1858, p. 3.

once, when everything was fresh in the minds of the people'.[11] Unlike Baines's cautious and restrictive prediction of the arrangements that could be made to allow public access to the Town Hall, Fairbairn urged the Town Council to consider opening every day, 'during reasonable hours', without an entrance fee. He had identified several areas that might be appropriate for pictures to be hung, including corridors and landings, the principal hall, the courts and justices' room.[12] Although this piecemeal approach was more practical, it would ultimately undermine the project and result in the use of paintings and sculptures as ancillary adornments to the architecture rather than the discrete art gallery first envisioned. Fairbairn mentioned the Royal Museum and Public Library in Salford, opened in 1850, and the Manchester Free Library, opened in 1852, as examples from which they could learn.[13]

The proposal met with support from the Town Council, and as the idea began to circulate more widely, so too did more critical opinions. An unnamed reporter from the *Leeds Intelligencer* foregrounded the paper's strong support for the idea of an art gallery for the town, but questioned the wisdom of its proposed connection to the Town Hall and, by extension, Leeds Corporation:

> many will be greatly in doubt as to the expediency of making the Corporation masters over the public gallery of art. If the project is to be carried on, it must be supported with some enthusiasm, and it must be constantly supported; while it does not follow that the members of the Corporation are, nor will be, Art-lovers.[14]

As the reporter recognised, a formal association with Leeds Corporation and the Town Council risked an art gallery and collection contingent on a constantly changing group of elected representatives, whose support for the scheme could not be guaranteed in perpetuity: a problem that continues to haunt local authority museums and galleries. Although a modest number of paintings and sculptures were donated for display in the Town Hall from its opening in 1858 onwards, it is clear that the public art gallery envisioned by the original group was not realised.

Just over a year after the opening of the Town Hall, members and supporters of the Leeds School of Art gathered at the Albion Street Music Hall to celebrate the prizes awarded to local pupils by the Department

[11] 'Proposed Fine Art Gallery for Leeds', p. 3.
[12] 'Proposed Fine Art Gallery for Leeds', p. 3.
[13] 'Proposed Fine Art Gallery for Leeds', p. 3.
[14] 'Art Gallery in Leeds', *Leeds Intelligencer*, 14 August 1858, p. 4.

of Science and Art. While they recognised the progress made in art education in Leeds, the speakers lamented the continued absence of an art gallery in the town and argued that the time was right to build a joint gallery, mechanics' institution and school of art, as the latter still subsisted in a garret at 22 East Parade in connection with the Leeds Mechanics' Institution and Literary Society. The new mayor, William Kelsall, continued his predecessor's rhetoric when he opened the proceedings with the declaration: 'whoever cultivated a taste for art and science would have his intellectual knowledge improved, and at the same time become a better citizen and a happier man'.[15] The chairman of Leeds School of Art, Peter O'Callaghan, gave a spirited speech which emphasised the great urgency of the situation and appealed to his audience's 'public spirit ... love of art, and, above all, on your love of your native town of Leeds' for their support.[16] These sentiments were echoed and extended by the president of the Leeds School of Art, the banker William Beckett Denison (1826–90), who encouraged his fellow townsmen to 'take advantage of the present prosperous state of trade, and abundance of money', in order that 'three years hence they might have the pleasure of congratulating themselves upon the erection of a commodious and magnificent building'.[17] His colleague James Hole reiterated the prospect of donations that were ready and waiting for a suitable home (which in his view did not include the Town Hall) and gave the opinion that 'the wealthier classes of Leeds should provide the means for erecting a commodious building in this town'.[18] Written in his capacity as honorary secretary of the Yorkshire Union of Mechanics' Institutes, Hole summarised the arguments presented at this meeting in his book on the state of working-class education in Leeds, published the following year in 1860, in which he argued:

> Along with proper class rooms [for Leeds School of Art], there should be combined an art gallery, where students might not only study from the best models, statuary, and pictures, but which might in the evening prove a pleasant resort for the inhabitants of the town, calculated to refine and elevate perceptions, to diffuse a love of art, and a taste for the beautiful. Experience would lead us to expect that, if a fitting receptacle were provided, the munificence

15 'Leeds School of Art: Proposed Art Gallery in Leeds', *Leeds Intelligencer*, 24 December 1859, p. 6.
16 'Leeds School of Art: Proposed Art Gallery in Leeds', p. 6.
17 'Leeds School of Art: Proposed Art Gallery in Leeds', p. 6.
18 'Leeds School of Art: Proposed Art Gallery in Leeds', p. 6.

> and public spirit of the community would fill it. Already some pictures have been given to the Corporation, and we know that several other gifts are merely waiting for a suitable place for their reception.[19]

Much like the Town Hall proposition, this modified suggestion relied on the conjunction of two or more functions in the same building, albeit more closely aligned in this instance. It was effectively an extension of the ways in which Leeds School of Art had operated since its establishment in 1846 and opening in 1847 as Leeds School of Design.[20] The school had staged public exhibitions, lectures and *conversazioni* from its inception and displayed its growing teaching collection as part of these events. The model of the amalgamated school-museum in this context was very much informed by its parent institution, the Government School of Design. Established at Somerset House in 1837, the reconstituted school moved to Marlborough House alongside the new Museum of Manufactures in 1852 before their combined transfer to South Kensington in 1857, as antecedents of the Royal College of Art and the Victoria and Albert Museum. The school-museum suited the objectives of the Department of Science and Art, the centralised governing body under the Board of Trade through which British art education was managed from 1853 onwards. In theory, it advanced its design reform agenda by educating both producers and consumers in the same space, while providing a more economical solution than constructing separate buildings.

While the Leeds Mechanics' Institution and Literary Society eventually succeeded in uniting its teaching activities under one purpose-built roof in 1868, it did not ultimately provide a home for a permanent art gallery. Although the discussions that took place in the late 1850s did not directly result in the solution so long hoped for, they were not futile, nor should they be considered failures. The advocacy for the Town Hall art gallery did result in some significant donations of paintings and

[19] James Hole, *"Light, more Light!" On the Present State of Education amongst the Working Classes of Leeds and How it Can Best be Improved* (London: Longman, Green, Longman and Roberts, 1860), pp. 70–71.

[20] Leeds School of Design was renamed the School of Ornamental Art in 1852 and Leeds School of Practical Art in 1853, although often referred to simply as 'Leeds School of Art' during this period. See Rebecca Wade, 'Pedagogic Objects: The Formation, Circulation and Exhibition of Teaching Collections for Art and Design Education in Leeds, 1837–1857' (unpublished doctoral thesis, University of Leeds, 2012).

sculptures to the town, now termed the Leeds Civic Collection.[21] It could be argued that the presentation of these art objects set an important precedent for this form of cultural philanthropy, which had until this point been limited to short-term loans to temporary exhibitions. Despite the desire for a permanent institution, these discussions also stimulated further temporary exhibitions from the 1860s to the 1880s.

The effect of the National Exhibition of Works of Art at Leeds

The first temporary public exhibition in Leeds to be organised with a stated objective to form a permanent art gallery for the town was the National Exhibition of Works of Art, as discussed in Chapter 5. Almost all the exhibitions we have considered so far resulted in published declarations in this direction *after* they had finished; the organisers of the National Exhibition of Works of Art announced their aims from the start. The communication of their plan was not limited to the local periodical press: it appeared in the *Illustrated London News* in March 1867, just over a year before the exhibition opened.[22] It was a bold strategy, but unfortunately it did not immediately culminate in the outcome the organisers had hoped for. Written five months after the exhibition's closure, a reporter for the *Leeds Times* recorded the aftermath in the following terms:

> It was hoped that the National Art Exhibition held in Leeds last year would not only do much good in the promotion of art culture amongst the classes who appreciated the grand display that was then thrown open to the public, but that it would be the means directly of giving to Leeds what the town has never hitherto possessed, but what it wants much—a permanent gallery of art. Perhaps in this respect the National Exhibition has not done all that it was expected to do; but something is being done, and the committee of the Mechanics' Institute, who have under their management one of the most successful of English art schools, are doing the work. The committee are filling this gallery with loan collections, and they were fortunate to obtain to begin with the

21 While nominally distinct from the wider collections of Leeds Museums and Galleries, both art collections are under the custodianship of Leeds City Council. Parts of the Leeds Civic Collection remain on display at Leeds Town Hall, while other works were transferred to Leeds Civic Hall after its completion in 1933.

22 'Fine Arts', *Illustrated London News*, 30 March 1867, p. 11.

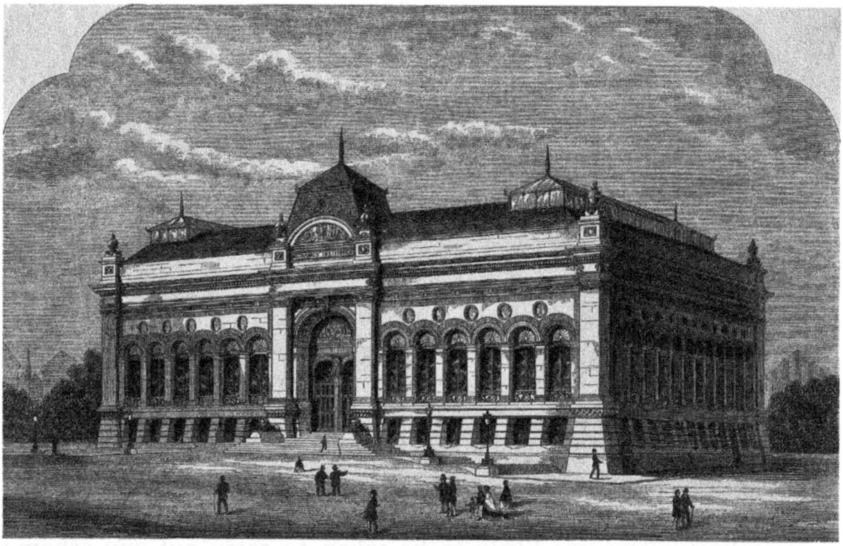

Figure 7.3 W.E. Hodgkin, *Leeds New Mechanics' Institution and School of Art*, from *The Illustrated London News*, 21 September 1867, wood engraving. Collection of the author

> valuable specimens of the old masters of the Italian, Spanish, and French schools shown at the Exhibition by Mr. Alexander Barker [and] a large collection of busts and statuettes from Mr. Arthur Marshall. The gallery will not be opened to the public until it is well filled, and meantime the committee solicit further contributions.[23]

It seemed for a time as though the hopes expressed a decade earlier might be realised, now that the Leeds Mechanics' Institution and Literary Society had a prominent building of its own on Cookridge Street, which also accommodated the Leeds School of Art (Figure 7.3). However, despite moderately successful loan exhibitions in 1869 and 1870, the Institute building was not to be the home of the sort of public art gallery envisioned by its promoters.[24]

23 'Proposed Fine Art Gallery in Leeds', *Leeds Times*, 20 March 1869, p. 5.

24 The Leeds Institute building would go on to become the home of Leeds City Museum in 2008, as the descendant of the collection formed by the Leeds Philosophical and Literary Society, exhibited at the Philosophical Hall until 1965.

The Yorkshire Fine Art Society

Following the pattern that erupted into the public discourse every decade or so, the call for a permanent art gallery for Leeds emerged again from hibernation in the late 1870s. Although the direction of the argument and much of the language remained consistent, the economic context had changed considerably over the course of the preceding ten years. Whereas the prominent voices in the later 1860s suggested that the time was right to take advantage of a period of prosperity, commentators at the end of the 1870s recognised the change in circumstances. In a letter written to the editor of the *Leeds Mercury* from 'W.T.' of Headingley on 5 November, the writer noted the problem but attempted to present a positive perspective:

> Of the desirability of having a permanent Art Gallery in Leeds there can be no question, but in the present depression of trade no hasty steps are likely to be taken to carry out any scheme for it. Perhaps it is as well, for time and leisure are thus afforded to consider how best it may be done when returning prosperity furnishes the necessary means.[25]

The writer proposed that, in the meantime, a temporary exhibition of the work of local artists could be held in the 'all but empty' picture gallery of the Mechanics' Institute. We might suspect that 'W.T.' was one of these local artists, when they called for means to be 'taken to assure them that they will be justly and fairly dealt with, and they are courteously invited to contribute rather than grudgingly and suspiciously permitted to do so'.[26] Although the writer did not give examples of instances when this had not been the case, the request does point towards the possibility that some artists who lived and worked in Leeds were not necessarily well-supported, or at least did not perceive themselves to be. It is worth noting that in this context the writer appeared to equate the term 'artist' with painters and disparaged those who worked in three dimensions: 'not that there is not local talent which uses the chisel rather than the brush or pencil for its expression, but in Leeds the sculptor is unfortunately a little too much mixed up with the stonemason for present discrimination'.[27] The implication here was that sculpture was too closely aligned with craft, trade and architectural

25 'Art Gallery for Leeds', *Leeds Mercury*, 9 November 1877, p. 2.

26 'Art Gallery for Leeds', p. 2.

27 'Art Gallery for Leeds', p. 2.

ornament for serious consideration as fine art; a prejudice common to Victorian art criticism. Although it may have been fair to say that Leeds was not well known for its sculptors during this period, local artists such as John Throp (1819–89) successfully negotiated the boundary between the fine and decorative arts. Throp combined architectural commissions in Leeds and beyond with the production of autonomous busts and figures. In addition to having regularly featured in exhibitions in Leeds and Liverpool, Throp's sculpture was considered highly enough to have been included in seven of the Royal Academy of Arts Summer Exhibitions between 1857 and 1880.[28]

As the writer of the above letter had hoped, momentum towards the establishment of a permanent art gallery began to build again. A small group first met in December 1876 to discuss the feasibility of appropriating some of the spaces used for the Yorkshire Exhibition of Arts and Manufactures at the Coloured Cloth Hall in 1875, only to be obstructed, 'as the Cloth Hall people declined to entertain it'.[29] On 30 October 1878 the group met at the Mechanics' Institute to consider the question anew, led by the then Vicar of Leeds, John Gott (1830–1906).[30] The record of the meeting published in the *Leeds Mercury* the following day noted that the group had hoped to instigate the construction of a permanent art gallery on land immediately behind the Mechanics' Institute building.[31] The proposal was abandoned after only £2,000 of the £10,000 guarantee fund was raised, with a small fraction of that amount derived from Leeds residents, whom the group criticised for their 'great lukewarmness' to the idea.[32] It was replaced with the intention to establish an art gallery in a temporary space. The brand new Assembly Rooms attached to Leeds Grand Theatre were considered as a possible venue and quickly discounted as too small.[33] The galleries of the Mechanics' Institute, used for exhibitions in 1869 and 1870, were rejected on the same grounds. The group found a solution in the as

28 'John Throp', *Mapping the Practice and Profession of Sculpture in Britain and Ireland 1851–1951*, University of Glasgow History of Art and HATII, online database 2011, http://sculpture.gla.ac.uk/view/person.php?id=msib5_1205412383 (accessed 19 January 2022).

29 'Proposed Fine Art Exhibition in Leeds', *Leeds Mercury*, 30 May 1879, p. 6.

30 'Proposed Fine Art Exhibition in Leeds', p. 6.

31 The land behind the Mechanics' Institute would eventually be used for the purpose of art education. The new Leeds Boys' Modern School building was completed in 1888 and the purpose-built Leeds School of Art in 1903. Both buildings now form part of the campus of Leeds Arts University.

32 'Proposed Fine Art Exhibition in Leeds', p. 6.

33 Leeds Grand Theatre opened on 18 November 1878.

yet unfinished Athenæum Buildings on Park Lane, designed by the architect William Bakewell (1839–1925). Having spoken with Bakewell about the suitability of the building for their purposes, a sub-committee of four members of the group was appointed to further consider the question, based on a rent of £250 per year for three years, with the option to extend the lease on the same terms for a further two years.[34] The proposal was agreed seven months later in a meeting at Leeds Town Hall on 29 May 1879, with the rooms expected to be ready by 1 November 1879.[35]

The Athenæum Buildings offered accommodation over two floors. The very top floor was approximately 23 metres long, 16 metres wide and 5 metres high and lit from the ceiling—then considered ideal for a picture gallery—with the floor below of the same size and lit from windows facing the street. Access to cellars for the storage of packing cases and a hoist to bring works of art (and visitors) to the upper floors made the arrangements more convenient than previous spaces appropriated for temporary exhibitions in buildings that had not been designed for the purpose.[36] Despite its many benefits, the group were not yet unanimous in their support for the use of the Athenæum Buildings, with Lancelot Wedderburn (a lender to the Yorkshire Exhibition of Arts and Manufactures in 1875 and discussed in Chapter 6) having offered an unspecified thirteen-room property of his own for the purpose, which was not thought suitable.[37] Another prominent building in the process of construction caught the attention of an unidentified correspondent to the *Leeds Mercury*, whose letter to the editor dated 22 September 1879 pointed towards what would become the Municipal Buildings as an appropriate candidate for the inclusion of an art gallery. The letter writer, who referred to themselves only with the initials 'J.F.C.', reinforced their argument with reference to the Art Gallery and Museum recently established as part of the Central Library by the neighbouring Bradford Corporation.[38] Cultural competition was now very close to home.

In response to this letter, the painter and tannery owner John Joseph Willson (c. 1837–1903) alerted the writer to the formation of the Yorkshire Fine Art Society, of which he was honorary secretary. Willson announced the Society's intention to sign a lease for the Athenæum

34 'The Proposed Permanent Fine Art Exhibition', *Leeds Mercury*, 31 October 1878, p. 5.

35 Untitled, *Leeds Mercury*, 30 May 1879, p. 4.

36 'Proposed Fine Art Exhibition in Leeds', *Leeds Mercury*, 30 May 1879, p. 6.

37 'Proposed Fine Art Exhibition in Leeds', p. 6.

38 'An Art Gallery for Leeds', *Leeds Mercury*, 23 September 1879, p. 8.

Buildings as soon as the guarantee fund reached £5,000, and concurred with the writer's points: 'I need not say I quite agree with him that Leeds has been left behind in the fine art race; and it is satisfactory to know that something is now being done to retrieve the past neglect.'[39] By the end of 1879 the Yorkshire Fine Art Society had consolidated its intentions with the formal election of an executive committee and the preparation of a set of rules. The guarantee fund stood just over its target at £5,240, and so the Society was now in a position to agree the lease on the Athenæum Buildings and plan the inaugural exhibition.[40]

At the beginning of 1880 the façade of the Athenæum Buildings was revealed as the scaffolding was dismantled. The *Leeds Mercury* published an assessment of the new building and a preview of its intended functions. The article noted that the Flemish-style façade stood 'in striking contrast to the dark-looking and smoke-begrimmed buildings adjoining' and commended its combination of beauty and utility as being 'as harmonious in general effect as it is pleasing in the details'.[41] In addition to the two floors leased by the Yorkshire Fine Art Society, the building was intended to accommodate a 'Travellers' Club'—for visiting barristers and other gentlemen visiting Leeds—and 'Ladies' Club', with four shops on the ground floor, including a restaurant and billiard room. Both the construction and decoration of the building were considered particularly suitable for the purpose of an art gallery because the floors were of Dennett's patent fireproof concrete and the external and internal ornament communicated a cultured opulence. American walnut doors, scagliola walls and mahogany window frames were coupled with carved stone pilasters, panels, spandrils and statues by local sculptor John Throp, which represented poetry, music and painting.[42]

With the building due for completion in March 1880, the first exhibition was planned to open on 1 June of the same year. The Society anticipated that around 800 works would be selected for display, including work by the 'principal local artists'.[43] The Society warned that the cost of staging the exhibition would be high, and that if they were to continue, the number of subscribers would have to increase. In addition to the subscription model of funding, the Society hoped to meet its expenses through commission on the sale of works of art and through

[39] 'An Art Gallery for Leeds', *Leeds Mercury*, 25 September 1879, p. 3.

[40] 'Yorkshire Fine Art Society', *Leeds Mercury*, 30 October 1879, p. 3.

[41] 'The Athenæum Buildings, Leeds', *Leeds Mercury*, 7 January 1880, p. 8.

[42] 'The Athenæum Buildings, Leeds', p. 8.

[43] 'The Fine Art Exhibition at Leeds', *Leeds Mercury*, 17 May 1880, p. 8.

ticket fees. The price of the tickets was to be differentiated at 1s., 6d. and 3d., with the cheapest admission only available to purchase on Tuesday evenings and Saturday afternoons and evenings.[44] The Society was keen to encourage working people to attend the exhibition to support its educative mission, although this appears not to have extended to the integration of the social classes.[45]

The first exhibition opened as planned on 1 June 1880. The president of the Society, George Robinson, 1st Marquess of Ripon (1827–1909), had been due to conduct the opening ceremony but this plan was disrupted by his appointment as Viceroy of India. He was replaced by Sir Philip Cunliffe-Owen (1828–94), a distinguished exhibition organiser with a longstanding involvement with the Department of Science and Art. He had been appointed the second director of the South Kensington Museum after Henry Cole's retirement in 1873.[46] In addition to this prestigious guest, the opening was attended by what the *York Herald* described as 'a large and influential company' composed of the local social, economic and political elite.[47] The South Kensington Museum was also represented by a loan collection of watercolours in addition to more than 1,300 pictures hung by the artists Richard Waller and William Howgate. Cunliffe-Owen praised the display in the following terms:

> The exhibition did great credit to those spirited men who had so admirably arranged the pictures. They might, he thought, be quite certain that it would lead before long to the establishment of a permanent gallery, which would not only encourage the artists of Leeds but would tend to elevate the whole population of that Mechanics' Institute.[48]

These sentiments reinforced the function of the exhibition as a temporary staging post on the way to a permanent art gallery. However certain these statements appear, though, it must be borne in mind that similarly definitive pronouncements had been made in Leeds for at least the previous forty years. Knowing in retrospect that Leeds was to have a purpose-built art gallery within eight years might give more

44 'The Fine Art Exhibition at Leeds', p. 8.

45 On opening the exhibition, Sir Philip Cunliffe-Owen used the opportunity to recommend that it be opened free of charge on Mondays and Saturdays, although his recommendation appears not to have been implemented. See 'Leeds Fine Art Exhibition', *York Herald*, 2 June 1880, p. 6.

46 'Leeds Fine Art Exhibition', p. 6.

47 'Leeds Fine Art Exhibition', p. 6.

48 'Leeds Fine Art Exhibition', p. 6.

weight to predictions of imminent progress the closer they appear to the realisation of this object; however, the trajectory was neither linear nor inevitable.

One of the honorary secretaries of the Society, the noted naturalist James William Davis (1846–93), echoed Cunliffe-Owen and expressed his ambition for their activities: 'we may hope that our success as a private society may ultimately encourage the authorities of the town to take the matter in hand, and add another institution of beneficence and culture to those already existing in Leeds'.[49] We might ask why Leeds Corporation would be compelled to take over an operation that ran successfully in private hands, at no cost to the public purse. Perhaps the most the Society could demonstrate was evidence that the people of the town and its environs were interested in the visual arts.

The first exhibition was successful enough for the Society to repeat the exercise fairly swiftly and the second exhibition opened on 15 September 1880. While the first exhibition was intended to display loans (although work for sale by living artists was admitted), the second was to be entirely dedicated to work by contemporary artists, offered for sale. Both established and emerging artists were invited to show their work, which this time was hung on the line (rather than through the more crowded salon-style hang of the previous iteration) by the artist Thomas John Gullick (c. 1828–95). The *Leeds Mercury* noted that the pictures had all been installed in time for the press view, with the catalogue ready and available. Although these points might not seem particularly worthy of comment, we have seen that this level of professional organisation was not always present. Exhibitions did not always open on time, objects were not necessarily all present at the opening and the production of a catalogue was often delayed. The second exhibition was successful by other measures too: it had attracted approximately 40,000 visitors by its close on 1 January 1881 and around 190 paintings were sold, to the value of £2,300.[50] The exhibition also benefited from royal patronage in the form of a visit by the Duke of Edinburgh on 13 October 1880.[51] The Society planned to extend its activities in 1881 by holding three exhibitions: of furniture, of works loaned by artists and of works loaned by private collectors.[52]

At the annual meeting held on 9 February 1881 the report noted that the Society had expenses of £3,846 and debts of £771. Its income from

49 'Yorkshire Fine Art Exhibition', *Leeds Mercury*, 5 June 1880, p. 5.

50 'Close of the Yorkshire Fine Art Society Exhibition, Leeds', *Leeds Mercury*, 8 January 1881, p. 5.

51 'The Yorkshire Fine Art Society', *Leeds Mercury*, 7 February 1881, p. 8.

52 'The Yorkshire Fine Art Society', p. 8.

subscriptions was £3,502 and the total from commissions generated by the first two exhibitions was £203. Despite the troubling debt, Gott reassured his colleagues that they did not begin their work 'as a money making affair' and rather sought to promote art and 'refine the popular taste' through exhibitions. He further told the assembled group that they could not have expected the first two exhibitions to clear the 'heavy expenditure' incurred at the start of their work.[53] In his position as Vicar of Leeds, Gott was perhaps less inclined to measure success in purely economic terms. He was more invested in the provision of 'healthy and elevating amusement', particularly for 'the young men whom our trade attracted to situations in this large town'. Gott thought that the wealth generated in Leeds should be paid back to its people in the form of art, as wealth for its own sake was, in his words, 'apt to corrupt and sink into grossness'.[54] Gott's beliefs were not only informed by his Christianity, but also by his position as the third generation of one of the most significant families in the development of industrial Leeds. The son of the mill owner and art collector William Gott (1797–1863) and grandson of the cloth merchant and manufacturer Benjamin Gott (1762–1840), John Gott experienced both the privileges of inherited wealth and the civic and cultural responsibilities of his position. His father and grandfather were celebrated philanthropists who contributed to the formation and maintenance of voluntary societies, public buildings and educational institutions in Leeds, in addition to William Gott's patronage of artists and notable collection of sculpture, paintings and prints.[55]

John Gott's peers Arthur and Victor Marshall, themselves descended from another foundational industrialist John Marshall (1765–1845), continued the cultural philanthropy embedded in their family's public life and presented a collection of plaster casts after antique sculpture to the Society. Of this development, Gott was recorded to have said:

> It was encouraging to find some of the oldest families in Leeds thus assisting one of the youngest institutions, and to find the rich, who had art in every room in their houses, willing and pleased to contribute valuable specimens for those whose ordinary

[53] 'Yorkshire Fine Art Society', *Leeds Mercury*, 10 February 1881, p. 3.

[54] 'Yorkshire Fine Art Society', p. 3.

[55] See R.G. Wilson, 'Gott, Benjamin (1762–1840), cloth merchant and manufacturer', *Oxford Dictionary of National Biography* (Oxford University Press, 2004), https://www-oxforddnb-com.libproxy.york.ac.uk/view/10.1093/ref:odnb/9780198614128.001.0001/odnb-9780198614128-e-49967 (accessed 25 January 2022).

> surroundings were less elegant. If this example were followed by other families in Leeds, we might soon have a good permanent exhibition, in addition to special periodical displays to attract visitors.[56]

Although the Society did not form a permanent collection of the sort envisaged here, the plaster casts donated by the Marshalls went on to form part of the foundational collection of Leeds Art Gallery on its opening in 1888.[57] The collection reproduced a number of the same casts of classical statuary that had been brought together for the Leeds School of Design after its establishment in 1846.[58]

The third exhibition opened on 1 March 1881 and closed on 30 April of the same year. As planned, it departed from the first two exhibitions in its concentration on antique and contemporary furniture rather than fine art, although examples of metalwork, clocks, prints, paintings, arms and armour and 'curiosities of various kinds' were also included.[59] The display of contemporary furniture on the lower floor represented the productions of a number of local manufacturers, and the arrangement appears to have been innovative for an exhibition of this sort in Leeds. Each manufacturer was allocated its own compartment in which to showcase its furniture, and some of the makers chose to decorate their area with wallpapers, carpets, window blinds and stained glass. By furnishing the spaces as complete domestic rooms, the exhibition encouraged visitors to imagine the products in their own homes. This form was more readily associated with trade shows, international exhibitions and the didactic displays of the Museum of Manufactures, which sought to educate its audience on the distinction between true and false principles of design in order to inform their choices as consumers.[60] The improvement of taste was repeatedly stated as an aim of the Society, although this was the first exhibition through which it

56 'Yorkshire Fine Art Society', *Leeds Mercury*, 10 February 1881, p. 3.

57 For further details, see Rebecca Wade, *Domenico Brucciani and the Formatori of Nineteenth-Century Britain* (New York: Bloomsbury Academic, 2019), pp. 124–26.

58 See Wade, 'Pedagogic Objects'.

59 'Exhibition of Ancient and Modern Furniture at Leeds', *Leeds Times*, 5 March 1881, p. 2.

60 See, for example, Clive Wainwright, 'Principles True and False: Pugin and the Foundation of the Museum of Manufactures', *Burlington Magazine*, 136 (1994) 357–64; Suga Yasuko, 'Designing the Morality of Consumption: "Chamber of Horrors" at the Museum of Ornamental Art, 1852–53', *Design Issues*, 20 (2004), 43–56 (p. 43); and Lara Kriegel, *Grand Designs: Labor,*

attempted to influence the public through objects other than paintings and sculpture.

In addition to recalling the rhetoric of the design reform movement of the middle decades of the century, the exhibition also referred back to the history of exhibitions in Leeds through the inclusion of curiosities and objects that had been displayed in the town before. The current Mayor of Leeds, George Tatham (1815–92), lent nine panes of glass reputedly etched with messages by Charles I during his incarceration at Carisbrooke Castle in 1646.[61] These kinds of historical curios with dubious provenance had been a feature from the earliest polytechnic exhibitions in Leeds, alongside objects of genuine antiquarian interest such as the collection of Fairfaxiana from Farnley Hall, once again loaned for display.[62] Having been a prominent feature of the Leeds Public Exhibition of 1839, the most recent opportunity local people had to encounter these objects was through the Yorkshire Exhibition of Arts and Manufactures in 1875, when Ayscough Fawkes opened Farnley Hall to paying visitors, as we saw in the previous chapter.

The notices of the 1881 exhibition had been positive, but by the beginning of 1882 the financial position of the Society was precarious. Rather than having reduced as it had hoped, the debt against the Society had instead doubled over the course of the year to £1,500, with guarantees amounting to £6,459. Despite the participation of people such as the artist Richard Waller, who had direct involvement with the ill-fated Leeds Academy of Arts described in Chapter 4, the majority of the committee seemed to be either unaware or in denial that the Yorkshire Fine Art Society was following a very similar trajectory. Only Joseph Lupton, supported by two of his colleagues, was in favour of winding up the Society with immediate effect. The rest of the committee agreed to continue to hold exhibitions until 1 March 1883, which was the earliest date that the initial three-year lease on the two floors of the Athenæum Building was due to end.[63]

The activities of the Society largely slipped out of the local periodical press during the following year, perhaps through a sense of embarrassment and pity for their peers and for the seemingly hopeless

Empire, and the Museum in Victorian Culture (Durham, NC: Duke University Press, 2007).

61 The nine panes of glass are now in the collection of the Victoria and Albert Museum.

62 'Exhibition of Ancient and Modern Furniture at Leeds', *Leeds Times*, 5 March 1881, p. 2.

63 'Yorkshire (Leeds) Fine Art Society.—Meeting of the Guarantors', *Leeds Mercury*, 31 January 1882, p. 5.

position of the visual arts in the town. When March 1883 came, so too did the news that the Yorkshire Fine Art Society had called on its guarantors to clear its liabilities, which had doubled again in the intervening year. The *Leeds Mercury* recognised that 'it would have been an act of folly to have continued the exhibitions in the hope that the hitherto steadily increasing debt would by some fortunate chance be paid off'.[64] It could be argued that the Society had already operated as an act of folly for the past year, and an even less generous interpretation would suggest that it had begun as one. William Howgate, himself part of the committee, took something close to this view when he 'reminded the meeting of the unwelcome truth that all fine art exhibitions in Leeds for fifty years past have proved financial failures'.[65] As we have seen, this was far from the truth, although the defeatism is perhaps understandable. Howgate continued:

> It may be the fault of those who have promoted these exhibitions; it may be the fault of the public; but the fact itself is indisputable; and the experience of the Society, the career of which was practically closed yesterday, is only a repetition of that with which all preceding attempts to establish art exhibitions in Leeds have been attended.[66]

The sense of frustration was palpable and the committee seemed bewildered that what they considered to have been a series of excellent exhibitions had not resulted in success, although their only tangible failure was in financial terms. The mayor had offered a lifeline in the form of an exhibition at the new Municipal Buildings, which opened in 1884, but the committee of the Society chose not to risk the possibility that another exhibition would exacerbate its losses. They concluded: 'For the present we must be content to accept the lesson which the experience of the Society teaches, and wait until there is a more genuine appreciation of fine art amongst the masses of the population of Leeds.'[67] Despite Howgate having conceded that the Society itself might have been to blame for its dissolution, it is clear from this summation that the primary target for his disappointment were the people of the town. It seems churlish to blame the 'masses of the population' when many of the individuals on the committee (and certainly their personal and professional networks) had the wealth to either completely

64 Untitled, *Leeds Mercury*, 15 March 1883, p. 4.
65 Untitled, p. 4.
66 Untitled, p. 4.
67 Untitled, p. 4.

fund or heavily subsidise their activities. It was the underlying belief that they should be self-supporting that undermined their progress. Perhaps what Leeds really failed to understand during this period was that exhibitions, art collections, galleries and museums tended (and continue to tend) to cost money rather than make money.

The state of the arts in Leeds was subject to national discussion in early 1884. The London-based *Magazine of Art* published an extensive article on the disparity between private collections and public institutions in the town.[68] The unnamed author pointed towards the widespread interest in 'Projects for the amelioration of the condition of the people', including the provision of free libraries, music and art galleries, and claimed: 'The attitude of the people towards art is no long one of stolid indifference.'[69] The era of the mechanics' institute was over, they wrote, because general standards of education had risen 'to a pitch never contemplated in those heathenish days, and are proceeding to higher things'.[70] Those higher things included fine art and an increasing demand for discrete public exhibitions of fine art, separated from displays of industry; such exhibitions, they argued, 'are more profitably dissociated from looms and cotton-mills and engines, and established in permanency apart from such'.[71] The author was incredulous that an important collector of modern pictures, A.S. Dixon of Leeds, could show their collection in neighbouring Huddersfield and Bradford but find no home for it in their own town. For the writer of this article, the prospect of a permanent art gallery in Leeds represented a cleansing and generative possibility, 'like sunshine in a place whose shades are Plutonian... Here may the sun of art diffuse a wholesome and joyous radiance, penetrate the ferruginous air and transform the gloomy tracts of ceaseless toil.'[72] The precise mechanism for this figurative transformation was not articulated here, except in the nebulous moral and educational terms that had become so common.

Along similar lines, a correspondent for the *Artist* journal—also published in London—wrote a short article on the position of the visual arts in Leeds at this time, which they described as being in 'as bad a state as it is possible to conceive'.[73] The writer cited the failure of unnamed recent picture exhibitions, an unclaimed donation of

[68] 'Pictures at Leeds', *Magazine of Art*, January 1884, pp. 238–45.
[69] 'Pictures at Leeds', p. 238.
[70] 'Pictures at Leeds', p. 238.
[71] 'Pictures at Leeds', p. 244.
[72] 'Pictures at Leeds', p. 245.
[73] 'Leeds', *Artist and Journal of Home Culture*, 1 February 1884, pp. 43–44.

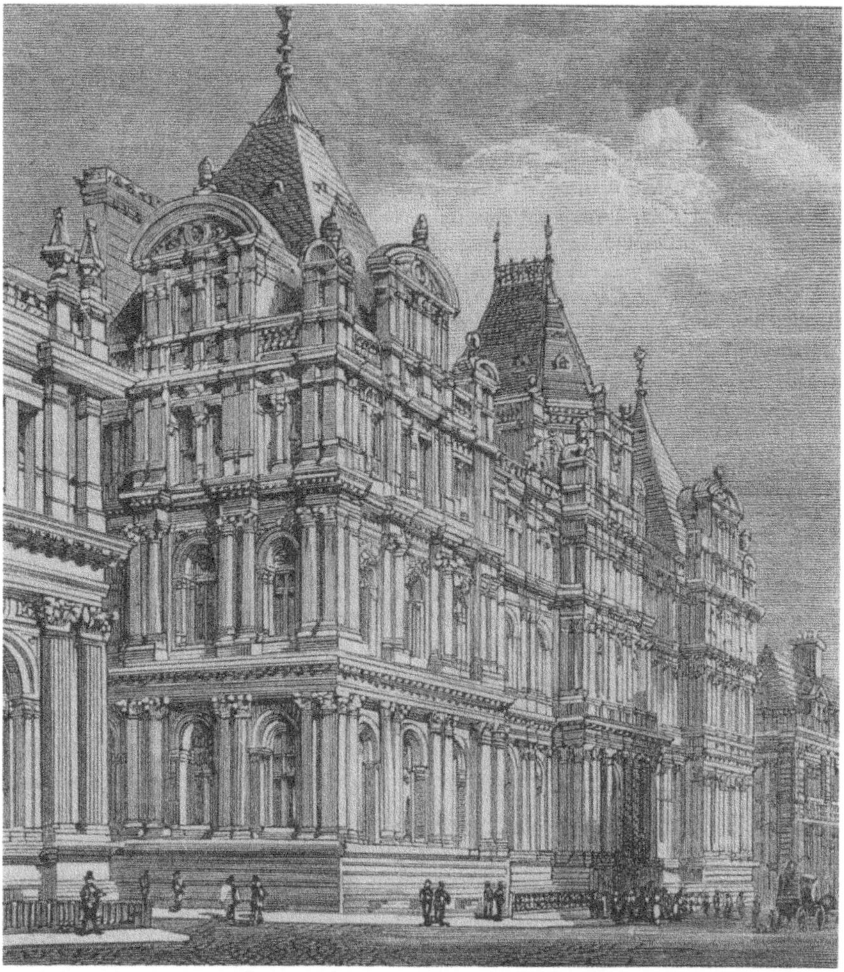

Figure 7.4 Anonymous, *The New Municipal Buildings and Free Public Library at Leeds: The New Building*, from *The Graphic*, 26 April 1884, wood engraving. Collection of the author

art and antiquarian objects to the town, the lack of a permanent art gallery and the inaccessibility of the few paintings and sculptures in the Town Hall as evidence for their argument. Leeds was again compared unfavourably with Liverpool, 'or even towns having scarcely half the population of Leeds!'[74] The writer expressed some qualified and cautious

[74] 'Leeds', pp. 43–44.

notes of optimism in the 'very fine' architecture of the new Municipal Buildings, the forthcoming exhibition of work by the members of the Leeds Society of Artists, the activities of the Leeds Photographic Society and an amateur sketching club in association with Leeds School of Art, which was judged to have been, 'on the whole, better than previous exhibitions'.[75]

Just as the opening of Leeds Town Hall in 1858 had generated discussion about the possibility of the new building accommodating an art gallery, the opening of the adjacent Municipal Buildings in 1884 had much the same effect (Figure 7.4). Among other civic functions, the Municipal Buildings accommodated a Free Public Library, and it was the Free Library Committee of Leeds Town Council that held a meeting on 4 September 1884 to discuss the question of the establishment of an art gallery in connection with the library.[76] A room alongside the lending department had already been set up as a small museum, containing six cases of loans from South Kensington for the period of one year, which included Sèvres ceramics, electrotypes, glass and metalwork, wood carving, pottery, textiles, coloured photographs and paintings by Dutch and Flemish old masters from the art collector William Frederick Barton Massey-Mainwaring (1845–1907).[77] As we have already witnessed in neighbouring Bradford, it was common from the second half of the nineteenth century onwards for libraries, museums and art galleries to occupy the same building as a result of legislation passed in the middle of the century, which allowed larger town councils to establish their own institutions through local taxation.[78] Therefore this modest but significant display in 1884 could be considered the first instance of an exhibition in Leeds having been staged in a truly public building and funded through local government.

In order to extend their activities, the Free Library Committee proposed to use a suite of seven rooms on the top floor of the Municipal Buildings, to be overseen by Edmund Bates (c. 1816–96). Bates was a local picture dealer, writer on art and early supporter of the artist John Atkinson Grimshaw (1836–93), tasked in this instance with procuring loans of pictures from local collectors. The disadvantage of locating the galleries so far from the ground floor was recognised at this preliminary stage, although the advantages of 'good light and perfect isolation' were

75 'Leeds', pp. 43–44.
76 'Proposed Art Gallery for Leeds', *Leeds Mercury*, 5 September 1884, p. 5.
77 'Proposed Art Gallery for Leeds', p. 5.
78 Principally the Museums Act of 1845 and the Public Libraries and Museums Acts of 1850 and 1855.

promoted as sufficient justification.[79] It was hoped that the ensuing exhibition would be ready to open by 16 October 1884.

Before the exhibition opened, John Lee, the writer of the regular 'Notes by the Owl' feature for the *Leeds Times*, saw the significance of the development: 'Readers of this column do not need to be told how sedulously the establishment of a museum and art gallery in Leeds has been advocated in the Leeds Times. Very pleasing is it to know that the Leeds Corporate authorities are moving in the right direction.'[80] Lee positioned the exhibition as the temporary nucleus of a permanent art gallery and one that committed the Town Council to finding larger and more accessible accommodation in the future, 'as Leeds does not like to play "second fiddle" to any other town in the country'.[81] He expressed confidence that the endeavour would be 'fully appreciated by the public of Leeds', but 'if the visitors should not prove so numerous at first as anticipated, the weary climb to such a great height will have to be held responsible for it'.[82]

Once the exhibition had opened, Lee took the opportunity to reiterate his views on the points both in favour of and against the scheme:

> Those of us who have striven for the establishment of a Free Fine Art Gallery and Museum in Leeds will rejoice that the thin edge of the wedge has been driven and that the future art prospects of this borough are flatteringly hopeful. We are not of those who fawn on the Corporation and praise all that our municipal legislators attempt, but in this instance Ald. Lupton and the Free Library Committee have laid the town under a deep debt of obligation by opening a fine art exhibition—which, though unfortunately placed too near the sky—possesses a high degree of merit. The rooms are not suitable for the purpose to which they are devoted. The lighting is defective, and there are other drawbacks, still the display is meritorious not only to the Corporation, but to every one concerned, and especially to those who have contributed selections from their art treasures.[83]

For Lee and other supporters of the venture, the exhibition represented irreversible progress towards their longstanding ambition to see Leeds

79 'Proposed Art Gallery for Leeds', p. 5.
80 John Lee, 'Notes by the Owl', *Leeds Times*, 6 September 1884, p. 8.
81 Lee, 'Notes by the Owl', p. 8.
82 Lee, 'Notes by the Owl', p. 8.
83 Lee, 'Notes by the Owl', p. 8.

construct a purpose-built, permanent, public art gallery. The source of funding and the building may have been new, but the contents of the exhibition were a familiar combination of loans from local individuals and South Kensington, which did not yet unite the idea of a public art gallery with the nascent public collection still housed at the Town Hall. The way forward was still unclear.

At the opening of another exhibition at the Municipal Buildings a year later, Charles Hargrove (1840–1918)—the minister at the Unitarian Mill Hill Chapel, with an active involvement in many of the town's intellectual, educational and charitable groups—used the occasion to establish a direct lineage between the present exhibition and the efforts of the Yorkshire Fine Art Society.[84] He erroneously claimed that the Society 'first began the movement in Leeds for public exhibitions of pictures and works of art', although he had only arrived in Leeds towards the end of 1876, so this misapprehension could be explained by either ignorance, exaggeration or a combination of the two. Hargrove acknowledged that the Society had not been successful in financial terms, but asserted that its work had been meaningful and influential:

> They knew that what they had done was a pecuniary failure, but they were quite sure they had done something to lay a foundation in Leeds for those who should come after them, and undertake the work under better auspices. Much sooner than they could have expected had this work been done by the Town Council, and to-day, he might say, was the triumph of the Yorkshire Fine Art Society, though it was after its death.[85]

The foundations for the Art Gallery and Museum at the Municipal Buildings had been laid by successive temporary exhibitions over the course of almost the whole century, and recent developments could be attributed to the confluence of more favourable social, political and economic circumstances.

In early 1886 John Lee of the *Leeds Times* proclaimed: 'A permanent art gallery for Leeds is in the near future.'[86] The evidence he provided for this confident declaration was in the strong attendance of 'many thousand persons' at the exhibitions held in 'the Leeds Municipal Palace', even though 'some 70 or 80 steps had to be ascended to reach ...

[84] 'Opening of the Leeds Art Exhibition', *Leeds Mercury*, 9 November 1885, p. 5.
[85] 'Opening of the Leeds Art Exhibition', p. 5.
[86] John Lee, 'Notes by the Owl', *Leeds Times*, 27 February 1886, p. 8.

this very high art gallery'.[87] The Museum and Arts Committee planned to hold another exhibition in November 1886, with an expanded remit to show work by artists from across the United Kingdom, rather than being limited to Yorkshire artists as represented by the previous exhibition. Towards the end of the article Lee gave a more moderate assessment of the conditions that could result in the construction of a new art gallery for Leeds:

> When a taste for the fine arts amongst our townsmen generally has had time to mature, the Corporation may have a little more money to spend—or some rich townsman may leave a lump, as we have heard of in other places—and then we shall have a permanent art gallery, and, let us live in hopes, a more easily accessible and appropriate gallery to show the genius of our artists to the best advantage.[88]

As this extract illustrates, the primary impediment remained a lack of funding, although there remained a tendency to hold a vague and undefined public responsible for the situation, through a lack of interest and taste that a permanent art gallery and collection were intended to stimulate. There still appeared to be a hope that Leeds would find its own Andrew Barclay Walker (1824–93), the brewer who had donated £20,000 to build the Walker Art Gallery for Liverpool a decade earlier.

By this point the rooms on the top floor of the Municipal Buildings had been used as a makeshift art gallery and museum for over a year, and there were few signs that this situation would alter without a significant change in circumstances. But rather than waiting passively for a substantial shift to occur, the second-generation industrial pin manufacturer Thomas Walter Harding (1843–1927) saw an opportunity to offer a solution that would at last result in a permanent, purpose-built art gallery for Leeds.

Thomas Walter Harding

Thomas Walter Harding looms large in accounts of the establishment and early years of Leeds Art Gallery. It is important, however, to balance the weight of his undoubted influence with other factors, to avoid attributing the whole scheme to one individual, while attempting

[87] Lee, 'Notes by the Owl', p. 8.
[88] Lee, 'Notes by the Owl', p. 8.

neither to overstate nor diminish his contribution to the cultural life of the town. Harding first entered the debate at a meeting of the Town Council on 8 April 1886, having been elected to represent the Headingley ward as a Liberal in 1883.[89] He moved: 'That in the opinion of the Council it is desirable that a building should be provided for the purposes of a permanent Fine Art Gallery and Museum for the borough, and that a committee be appointed to consider the subject and report thereon to the Council.'[90] There is very little to distinguish this motion from those that had gone before, except that Harding followed it with a impassioned speech that articulated not only the familiar reasons in favour of the idea, but also a detailed and practical proposal for how it could be brought into being. He gave the customary argument that European arts and manufactures remained superior because the state prioritised the provision of public art galleries and museums, where working people encountered exemplary works of the fine and decorative arts. In addition to the improvement of manufactures, Harding presented art galleries as sites of 'refined enjoyment' that gave respite to working people from their otherwise 'dismal' lives and 'the monotonous dullness, the weariness of the struggle for existence when so many had to carry on, often in the midst of the most unlovely surroundings and in great temptations'.[91] Here was the paternal impulse of the industrialist to direct the moral and social lives of his workers towards ever greater productivity.

The next instalment of his argument deployed the shibboleth of regional civic rivalry. Birmingham, Liverpool, Manchester and 'even their neighbours of Bradford' already had their own public art galleries, and Harding hardly needed to mention the longstanding chagrin this had caused among his peers, embarrassed by Leeds's deficit.[92] Harding did not only concentrate on what was lacking, however: he recognised what had already been achieved in the town. He praised the 'very fine' library and encouraged his colleagues to see their work as complete in

89 Melanie Hall, 'Harding, (Thomas) Walter (1843–1927), pin manufacturer and civic art patron', *Oxford Dictionary of National Biography* (Oxford University Press, 2004), https://www-oxforddnb-com.libproxy.york.ac.uk/view/10.1093/ref:odnb/9780198614128.001.0001/odnb-9780198614128-e-96377 (accessed 28 January 2022).

90 'Proposed Art Gallery and Museum for Leeds', *Leeds Mercury*, 10 April 1886, p. 4.

91 'Proposed Art Gallery and Museum for Leeds', p. 4.

92 'Proposed Art Gallery and Museum for Leeds', p. 4. On the civic rivalry between Leeds and Bradford, see Asa Briggs, 'Leeds: A Study in Civic Pride', in *Victorian Cities* (London: Odhams, 1963), pp. 139–83.

this area. Harding acknowledged the efforts of the Library Committee in the establishment of art exhibitions alongside their literary activities, but echoed the most common criticism: 'It required a great amount of enthusiasm to mount a hundred or a hundred and twenty steps to see a collection of pictures... If they were to have a permanent Art Gallery one thing was quite certain—it must not be in the clouds.'[93] For Harding it was crucial that the proposed art gallery be accessible and purpose-built, in order to convey favourable messages about the town to both its inhabitants and its competitors.

At this point in the proceedings Harding shifted from abstract arguments to concrete plans. He pointed out that there was little building work in progress in the town, 'and consequently they would get the work done more cheaply than in more prosperous times'.[94] He suggested that a secondary consequence of construction was the employment of 'a great many people who were now in distress', through which he displayed a utilitarian approach to the economic condition of unemployed people in the town. To reduce costs further, Harding proposed that the new art gallery should adjoin the Municipal Buildings and be constructed on vacant land to the rear of the existing building, which was already owned by Leeds Corporation. Where Harding went much further than any previous proposal was in his preparation of architectural plans and estimates. For approximately £10,000, he contended, they could build a two-storey art gallery to include a sculpture gallery with room for museum and loan collections, three top-lit picture galleries and a suite of ten small cabinet rooms, 'after the fashion of Munich and Dresden'.[95] Harding initially put forward the idea of borrowing the £10,000 at a projected interest rate of 3.5%. With the inclusion of a 1% sinking fund, he sought to tempt the Town Council into action with the prospect that they could instigate the scheme for only £500. He implored: 'Could they not reduce the expenditure on the library and do something for art?'[96] Harding attempted to close the deal by pointing towards the losses suffered by the town through its lack of a permanent art gallery and the gains to be generated by its expeditious establishment. He speculated that Leeds could have been the recipient of the large and important collection formed by the local cloth manufacturer John Sheepshanks (1787–1863), which Sheepshanks

93 'Proposed Art Gallery and Museum for Leeds', p. 4.
94 'Proposed Art Gallery and Museum for Leeds', p. 4.
95 'Proposed Art Gallery and Museum for Leeds', p. 4.
96 'Proposed Art Gallery and Museum for Leeds', p. 4.

instead presented to the South Kensington Museum in 1857.[97] This may have been the source of persistent claims that Sheepshanks had either intended to donate his collection to Leeds, or had indeed presented it only for it to be rejected.

The Town Council appeared broadly to support Harding's arguments, although Alderman Frederick Robert Spark (1831–1919) suggested that Harding's resolution be diluted to support the appointment of a committee 'to consider the desirability and practicability of providing an gallery and museum', to report to the Town Council.[98] The amendment suggested that the council did not already consider the provision of an art gallery as inherently desirable, as Harding had moved at the start of the meeting. Despite this minor impediment, the resolution was carried by a large majority and a committee was appointed, which included Harding.[99]

Leeds Corporation's tentative agreement to engage with the proposal for an art gallery operated in the context of a broader 'municipal socialism', which Simon Gunn has argued reached its zenith in the 1870s and 1880s.[100] Like other comparable incorporated industrial towns, Leeds had brought the provision of certain services and 'natural monopolies' such as water and gas under the control of local government. The conditions were more conducive to the extension of intervention into the sphere of culture than they had been during the previous generation. The increase in activity came with concomitant costs, so it was crucial to secure the general consent of ratepayers for any major spending or diversion of resources. As the historian Martin Daunton has articulated in his study of taxation and representation in this context, urban governance depended on the construction and maintenance of a balance between individual and social costs and benefits.[101] To secure wide enough support for a particular scheme, it

97 Richard Redgrave, *On the Gift of the Sheepshanks Collection, with a view to the Formation of a National Gallery of British Art* (London: Chapman and Hall, 1857).

98 'Proposed Art Gallery and Museum for Leeds', p. 4.

99 The committee was composed of the following men: Mayor Gaunt, Aldermen Lupton, Boothroyd, Spark and Woodhouse and Councillors Harvey, Gilston, Ambler, Cooke, Lowe, Metcalf, Tweedale, Pollock, Wilson and Harding.

100 Gunn, 'Ritual and Civic Culture in the English Industrial City, *c.* 1835–1914', p. 237.

101 Martin Daunton, 'Taxation and Representation in the Victorian City', in *Cities of Ideas: Civil Society and Urban Governance in Britain 1800–2000*, ed. by Robert Colls and Richard Rodger (Aldershot: Ashgate, 2004), pp. 21–45 (p. 24).

was crucial that decision-making processes were transparent and made a matter of public record through newspaper coverage. As such, Harding participated in what Morris described as 'a culture in which knowledge, information and argument gained an impact in debate, decision and the allocation of power and legitimacy'.[102] Harding was aware that his persuasive rhetoric would reach the local electorate through the periodical press, but not without the mediation of the journalists whose endorsement could influence opinion.

The *Leeds Mercury* published a commentary on the proceedings in the same issue that reported Harding's proposal. The unnamed journalist expressed support for the original motion, but took issue with his first line of argument, that Leeds needed an art gallery to improve the 'dismal' lives of its working population:

> There is always some compensation, however hard the lot may be; and life in our large industrial centres, as compared with that in a quiet, dull, country parish, is not so barren of sources of enjoyment as in our pessimist moods we are sometimes disposed to believe and assert.[103]

The reporter further contended that if the intellectual life of the town was so impoverished, Leeds would not be aware of what it lacked and there would be no call for public art galleries, museums or libraries. They thought that the stronger argument in favour of an art gallery was educational, and, further, that this was the only grounds on which to support these institutions from public funds. On Harding's other points, the writer was in accordance and reiterated the call for a permanent art gallery to be well lit, simple in its design and located in an accessible, central position. In their opinion, art exhibitions in Leeds had generally failed, 'because the surroundings were uninviting and because the lighting has been defective'.[104] As we have seen, temporary exhibitions in Leeds had struggled with effective lighting as a result of their use of appropriated spaces, often lit by side-facing windows with a tendency to cast works of art either in deep shadow or excessive illumination.

Harding's speech also resulted in correspondence from readers of the *Leeds Mercury*, including a letter from James Reid of Chapel Allerton, then a village north of the centre of Leeds, who had published a book

[102] Morris, 'Governance: Two Centuries of Urban Growth', p. 8.
[103] Untitled, *Leeds Mercury*, 10 April 1886, p. 6.
[104] Untitled, p. 6.

on the moral teachings of academy paintings two years earlier.[105] Reid sensed the possibility of an opportunity to generate momentum and used his letter to the editor to urge 'some philanthropist to immortalise his name' on the model of the Walker Art Gallery in Liverpool.[106] He noted that opponents of the idea argued that the strong attendance at the two recent exhibitions in the Municipal Buildings was a result of 'pleasure-seekers from the surrounding districts', because 'the people of Leeds do not care for art'.[107] Reid challenged the legitimacy of both claims and countered with a position that remains familiar to this day:

> Vast numbers of people cannot surely be drawn to the town by the attraction of art treasures, &c., without leaving some trace of their visit in the receipts of some of the heavily rated shopkeepers. And, then, is there not a sense of pride in every true-hearted townsman in being able to give pleasure and benefit to others? To share our pleasures is to ennoble ourselves.[108]

There is much evidence that temporary exhibitions in Leeds had been visited by people from outside the town, with special trains organised for day trips from across the country. It was unusual, however, to see cultural tourism presented as a line of reasoning in support of a permanent art gallery for Leeds. It broadened the rationale of serving its own people to include an appeal to a civic pride that was generous to its neighbours. Reid pressed the familiar sore point: 'Towns of less importance and wealth are doing their duty in this respect, and the time has come when Leeds cannot longer afford to remain in the undignified position it sustains in relation to art matters.'[109] Unlike the journalist of the *Leeds Mercury*, Reid agreed with Harding on his support of an art gallery as a means to partially ameliorate the conditions in which the working classes lived and worked. He described the art gallery as 'a cheering antithesis to a cheerless back-to-back cottage, an occasional withdrawing room to the toil-worn labourer and his family, and a counteracting influence to the pursuits and occupations of a vitiating nature'.[110] Aside from their loftier educational aims, art galleries and museums could provide the respite of light, space and warmth to a

[105] James Reid, *Academy Paintings and their Moral Teaching* (Leeds: Richard Jackson, 1884).

[106] 'A Permanent Art Gallery for Leeds', *Leeds Mercury*, 16 April 1886, p. 5.

[107] 'A Permanent Art Gallery for Leeds', p. 5.

[108] 'A Permanent Art Gallery for Leeds', p. 5.

[109] 'A Permanent Art Gallery for Leeds', p. 5.

[110] 'A Permanent Art Gallery for Leeds', p. 5.

population who did not necessarily experience these qualities in their domestic or working lives.

The Fine Art Gallery and Museum Committee

Harding was appointed chairman of the new Fine Art Gallery and Museum Committee, which met for the first time on 19 April 1886. Several important plans were presented at this inaugural meeting, which would go on to define the ways in which the project was eventually realised. The proposal to construct the art gallery as an extension of the Municipal Buildings was reiterated, alongside a new idea to commandeer the library's existing reading room as a sculpture gallery and replace it with a new room on the ground floor of the art gallery, with its own entrance on the corner of Alexander Street and Centenary Street (Figure 7.5). The budget of £10,000 remained and it was suggested that £600 or £700 per annum be diverted from the Free Libraries as they were considered now to be 'tolerably well stocked'.[111] What Harding had presented as a preliminary plan based on estimates quickly consolidated into action, to the concern of some of his colleagues. The committee were increasingly called on to justify their decisions and consider alternative proposals. They defended their preference for a modest scheme by pointing out that, in their opinion, it would cost at least £50,000 to build a separate gallery in a central location with a more elaborate façade.[112] The local architect and member of the committee Thomas Ambler (1838–1920) submitted his own plan, estimated to cost even less at £8,000. Although Ambler's plan informed the process, particularly through his role on a sub-committee with Harding and fellow architect John Tweedale, he did not receive the commission.[113]

The Fine Art Gallery and Museum Committee presented their report to the Town Council in the summer of 1886.[114] Harding reiterated the points in its favour and promised that he would donate a 'considerable sum' to support the purchase of new acquisitions for the collection.[115] The committee's proposal to transfer £800 per annum from the penny rate intended for libraries was divisive and the Town Clerk was asked to

[111] 'The Proposed Corporation Fine Art Gallery and Museum in Leeds', *Leeds Mercury*, 20 April 1886, p. 5.

[112] 'Proposed Fine Art Gallery for Leeds', *Leeds Mercury*, 10 June 1886, p. 7.

[113] 'Proposed Fine Art Gallery for Leeds', p. 7.

[114] 'The Proposed Corporation Art Gallery and Museum for Leeds', *Leeds Mercury*, 31 July 1886, p. 10.

[115] 'The Proposed Corporation Art Gallery and Museum for Leeds', p. 10.

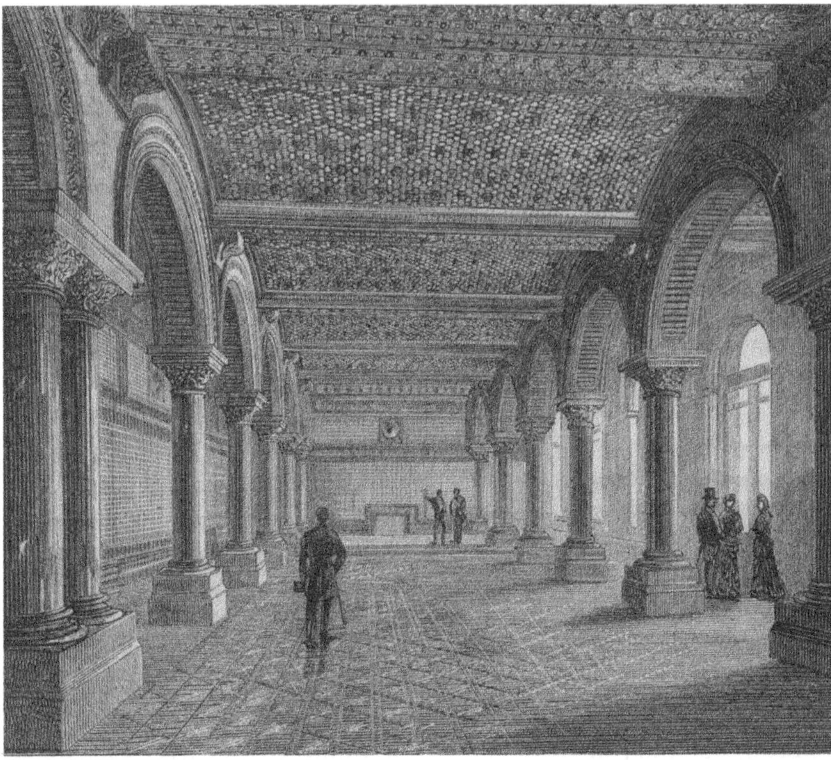

Figure 7.5 Anonymous, *The New Municipal Buildings and Free Public Library at Leeds: The Reading Room in the Free Public Library*, from *The Graphic*, 26 April 1884, wood engraving. Collection of the author

compile a report on how other towns funded their art galleries. It was found that most towns used the penny rate except in cases where an Act of Parliament had been requested to exceed this in special circumstances. The committee were reluctant to impose on ratepayers any further, fearing that they 'might not approve the erection of a costly building on a conspicuous site'.[116] The committee summarised their position on the proposed site as follows:

> The committee considered this site excellently adapted to the purpose, because of its central and quiet position, because its situation at the back of the Municipal Buildings rendered ornament

[116] 'The Proposed Corporation Art Gallery and Museum for Leeds', p. 10.

> and expense in the exterior of the new structure unnecessary, and because by approaching the new buildings through the existing news-room ... the art galleries would really form part of the Municipal Buildings, and be worthily approached through them.[117]

The extent of the integration of the art gallery into the Municipal Buildings was now made clear: not only would the art gallery be conjoined with the existing building, it would be entered through it and would therefore benefit from its symbolic and civic capital. Furthermore, the boundary between the library and the art gallery would be made permeable by the relocation of the reading room.

Although the Free Library Committee had already approved the plan before it was put before the Town Council, Harding faced some opposition over the reduction of the budget intended to maintain the twenty-five branch libraries in the town. Harding responded that the money raised under the Libraries and Museums Act would still amount to £4,600 per annum for the libraries if the art gallery was allocated £800 per annum. He declared: 'If he [Harding] had thought that the proposal would result in starving the library he would not have submitted the scheme.'[118] Alderman William Emsley objected to the proposal on the grounds that 'the museum and sculpture gallery were to be given the best part of the building, and that there was already a museum in existence in the town'.[119] Harding did not address the first concern over the annexation of the richly decorated reading room, but he did strongly counter Emsley's perception that Leeds did not require another museum. The existing museum referred to was the collection of the Leeds Philosophical and Literary Society, housed in part of its purpose-built Philosophical Hall on nearby Park Row. Harding stated that it was not their intention to duplicate existing provision, but rather to display 'articles of art manufactures and others from South Kensington'.[120] Although it was not referred to in the debate, there were crucial distinctions to be made between a collection established by a private society and a collection established through public funds. The implicit message was that the new museum and art gallery would be more accessible to working people, although the membership and governance of the Leeds Philosophical and Literary Society substantially

[117] 'The Proposed Corporation Art Gallery and Museum for Leeds', p. 10.

[118] 'A Permanent Art Gallery and Museum for Leeds', *Leeds Times*, 7 August 1886, p. 3.

[119] 'A Permanent Art Gallery and Museum for Leeds', p. 3.

[120] 'A Permanent Art Gallery and Museum for Leeds', p. 3.

intersected with the composition of the Town Council. As a result, the acquisition and interpretation of objects remained in the hands of broadly the same elite coterie.

The Town Council were not unanimous in their support for the scheme: twenty-one voted in favour and fourteen against.[121] As the plans became more concrete, they became more divisive. In his 'Notes by the Owl' column, the *Leeds Times* journalist John Lee expressed the opinion that the art gallery could occupy the site of a former cattle market on North Street, half a mile north-east of the Municipal Buildings.[122] Lee argued that the reading room was not appropriate for the display of sculpture, being too dark and lit from the side. The reading room was designed with a very large bank of windows on its south side; however, they immediately faced the buildings on the opposite side of the narrow Centenary Street, which were demolished between 1928 and 1932. It is important to note that the façade of the new art gallery would face these buildings too, with a limited view from the street. That the proposed site was surrounded on all sides was promoted by its supporters for its quiet position and, more pragmatically, because it obviated the need for an elaborate and expensive frontage. Lee also criticised this economising:

> Do not let us, for the sake of cheapness, adopt an idea for the Art Gallery and Museum that we shall be ashamed of in the future... We are very anxious, like our fellow townsmen, for an art gallery, but do not let us be hasty and then have to rue. Let us, by all means, have a building that will be a credit to Leeds.[123]

If it were a source of collective embarrassment that Leeds had not yet built its own art gallery, it was at least not in direct competition with those who had. Once Leeds had committed to a course of action, it was also committed to the possibility of unfavourable comparison with its rivals. How could Leeds win when it would only spend a fraction of what other towns had allocated? These discussions returned to the disparity between those who wanted the result as quickly as possible to capitalise on the momentum that had been generated, against those who were willing to wait for more favourable circumstances.

121 'A Permanent Art Gallery and Museum for Leeds', p. 3.
122 Now the site of Lovell Park.
123 John Lee, 'Notes by the Owl', *Leeds Times*, 14 August 1886, p. 8.

The architectural competition

The pragmatists seized the moment and by September 1886 the committee had instigated an architectural competition to design the new art gallery.[124] Originally restricted to architects who practised in Leeds, the competition was then opened to national entries with an extended deadline of 16 October 1886.[125] This decision antagonised members of the Leeds and West Yorkshire Architectural Society, who sent a letter of protest to the Leeds Town Clerk signed by twenty-three architectural firms, published in the *British Architect* journal.[126] The president of the Society, Charles R. Chorley, attacked what he described as

> the erratic and aimless movements of the Corporation Committee who had in charge the arrangements for the erection of an art gallery for Leeds, especially in rejecting, somewhat brusquely, the advice of that society regarding the sum to be expended on the new building. How absurdly ridiculous the amount fixed for the erection of the art galleries was in comparison with the work to be done any rational and honest architect would admit, when he heard that the rate per cubic foot which the Leeds Art Galleries Committee considered sufficient to erect public buildings hardly exceeded 3d. (Laughter). That, too, to include fire-proof floors, and all the special arrangements for lighting, ventilating, heating, &c., which the future presence of valuable contents demanded.[127]

Although the Fine Art Gallery and Museum Committee included two architects, Ambler and Tweedale, Chorley thought that the local architectural community had been slighted: 'The curt disregard which the committee has shown for skilled testimony on the question would certainly end in their landing themselves in a position from which they could only be withdrawn with ridicule to themselves and injustice to others.'[128] The opinion of one of his colleagues was even stronger: 'The building would be an unmitigated evil and a nuisance. People

124 'The Proposed Fine Art Gallery in Leeds', *Leeds Mercury*, 10 September 1886, p. 5.

125 'The Public Art Gallery for Leeds', *Leeds Mercury*, 17 September 1886, p. 5.

126 'Leeds Fine Art Gallery Competition', *British Architect*, 24 September 1886, p. 300.

127 'Leeds Architects and the Fine Art Gallery: A Fling at the Art Committee', *Leeds Times*, 23 October 1886, p. 3.

128 'Leeds Architects and the Fine Art Gallery: A Fling at the Art Committee', p. 3.

would never give pictures or other art products to be placed in the most beggarly public structure in Leeds.'[129] Their reservations were understandable, as the £10,000 projected budget for the art gallery was less that 10% of the cost of the Municipal Buildings to which it was to be attached.

Members of the local intellectual community expressed similar reservations. Nathan Bodington (1848–1911), then Principal and Professor of Greek at the Yorkshire College, gave a lecture on art museums to the Wortley Working Men's Institute in November 1886. With reference to the plans for Leeds, Bodington was reported to have commented:

> The object appeared to be to get a building for the smallest possible amount of money, and as to collections of art objects, he believed they were to take care of themselves. Honestly, he did not think that was a good policy. He was not fond of paying high rates; at the same time he did not think much of cheeseparing economy—and he did not think to do a thing of that kind poorly was in the true interest of the working men of Leeds.[130]

The thoughts of the working men to whom Bodington's opinions were addressed went unrecorded, but we might reasonably assume that they were not fond of paying high rates either. His advocacy was not directed towards art museums that only displayed paintings. Bodington hoped for an institution that would serve the needs of its local constituency, 'for in the centre of great textile works, such as we had in Leeds, we ought to have specimens of beautiful textile fabrics, in order that the operatives might be able to imitate them'.[131] The notion that operatives were afforded the autonomy to imitate exemplars seems naïve, and to have been inherited—as we have repeatedly encountered—from the design reform movement of the mid-century. In his concluding remarks Bodington appeared to concede that the members of his audience were unlikely to experience economic benefits from art knowledge gleaned through a gallery: 'even though the practise of art did not put a shilling in their pockets or contribute a farthing to the prosperity of the town or country, still its pursuit was a very high privilege. He could imagine no

[129] A comment by Mr L. Cooper in 'Leeds Architects and the Fine Art Gallery: A Fling at the Art Committee', p. 3.

[130] 'Wortley Working Men's Institute: Prof. Bodington on Art Museums', *Leeds Mercury*, 23 November 1886, p. 8.

[131] 'Wortley Working Men's Institute: Prof. Bodington on Art Museums', p. 8.

higher privilege than to acquire a cultivated eye and hand.'[132] Bodington was not alone in his Ruskinian idealism, which could be interpreted as patronising and detached from the realities of the lives of working people.

Despite the critics who wanted to see Leeds continue the architectural ambition of the Town Hall and neighbouring Municipal Buildings, the plans for a modest art gallery continued. Only days after Bodington had delivered his lecture, the competition to design the building was decided and the plans displayed for public view at the Town Hall. Competing architects had been given a strict brief and instructed that the cost of their realised designs was not to exceed £8,000.[133] Even though the competition had been opened up to entries from beyond Leeds, only five of the 18 submissions were not based in the town.[134] Of the local entries, some of the noteworthy architects included Charles Chorley (who had been so scathing about the scheme in his capacity as president of the Leeds and West Yorkshire Architectural Society), George Corson (designer of the Municipal Buildings, to which the art gallery was to be directly attached) and William Bakewell (designer of the Athenæum Building, former home of the Yorkshire Fine Art Society exhibitions).

The committee longlisted eight of the proposals and shortlisted three local architects: Fraser and Sugden of Park Square, George Corson of Cookridge Street and W.H. Thorp of Park Row. Thorp was awarded the first prize of £50 and Fraser and Sugden the second prize of £25. The reason for Corson's relegation to third place was not recorded. As the architect with the greatest claim on the prize—having designed the building to which the art gallery was to be integrated, in addition to other prominent public and private commissions in the town—we might speculate that the committee were perhaps uncertain of his ability to control his tendency to apply intricate architectural ornament and make use of expensive materials. The winning design by William Henry Thorp (1852–1944) adhered closely to Harding's original proposal. Described as Italian Renaissance in style, it was praised for its

[132] 'Wortley Working Men's Institute: Prof. Bodington on Art Museums', p. 8.

[133] 'Leeds Fine Art Gallery and Museum', *Leeds Mercury*, 27 November 1886, p. 10.

[134] The entries from Leeds were by Mr C.S. Nelson, Mr J.P. Key, Messrs Perkin and Bulmer, Messrs Dodgson and Braithwaite, Messrs Fraser and Sugden, Mr J.H. Greaves, Mr W.H. Thorp, Messrs Chorley and Connon, Mr George Corson, Mr J. Hall, Mr A. Neill, Mr W. Bakewell and Mr B. Wilson. Other entries were from Mr C.H. Mileham (London), Mr J. Johnson (London), Mr H.A. Cheers (Twickenham), Mr H.A. Matear (Southport) and Mr P. Ogden (Manchester).

harmonious relationship with the Municipal Buildings and its internal arrangement informed by the Walker Art Gallery and the 'more modern courts' of the South Kensington Museum.[135] Crucially, the estimate for its construction was £7,999 13s. 6d., although Thorp noted that for £10,000, 'greater satisfaction could have been given in the execution of the work'.[136]

The Fine Art Gallery and Museum Committee had made their selection, but Thorp's design was still subject to ratification by the Town Council. Before the council met to consider the proposal, those who had already objected to the principle of the plan took the opportunity to restate the terms of their disapproval. John Lee of the *Leeds Times* was the first to criticise the architectural plans displayed in the Law Library of the Town Hall:

> There is nothing very striking about the plans. They all impress one with the idea that they were designed to come within the £8,000 to which the architects were restricted, and hence several of the plans are bald in their plainness ... the plans will do; but, at the best, the whole scheme is hardly worthy of art or of the importance of Leeds. The architects were all of opinion that the sum fixed upon was too small.[137]

Lee further expressed resignation about the decision to hand the 'gaudy and darksome' reading room over to the art gallery for the display of sculpture, which he believed to be a fait accompli.[138] He was joined in this view by Professor Bodington, through a joint letter to the *Leeds Mercury* written with his colleague at the Yorkshire College, Louis Compton Miall (1842–1921), then Professor of Biology and curator of the museum of the Leeds Philosophical and Literary Society.[139] Bodington and Miall diverged from Lee in their broad support for the architectural plans, which they described as 'a skilful treatment of a difficult site'.[140] They continued to promote the idea that 'by cultivating the sense of beauty it may be possible to increase our power of industrial production' and urged those responsible for acquiring objects for the collection to resist the temptation to purchase or accept 'second-rate pictures', in

[135] 'Leeds Fine Art Gallery and Museum', p. 10.
[136] 'Leeds Fine Art Gallery and Museum', p. 10.
[137] John Lee, 'Notes by the Owl', *Leeds Times*, 4 December 1886, p. 8.
[138] Lee, 'Notes by the Owl', p. 8.
[139] N. Bodington and L.C. Miall, 'The New Municipal Art Gallery in Leeds', *Leeds Mercury*, 8 December 1886, p. 8.
[140] Bodington and Miall, 'The New Municipal Art Gallery in Leeds', p. 8.

order to fill the walls of the new art gallery quickly. Instead Bodington and Miall advocated a slow and deliberate approach to the development of the collection, focused on excellence and relevance to the industries of Leeds.[141] They presented Birmingham Museum and Art Gallery, opened during the previous year, as a model to emulate in this respect, in contrast to the exhibitions of the Yorkshire Fine Art Society, which they presented as a lesson in 'what happens when the managers have large rooms and are afraid to leave blanks'.[142]

The local painter John Joseph Willson, former honorary secretary of the Yorkshire Fine Art Society, wrote to the *Leeds Mercury* in agreement with Bodington and Miall. Willson thought that the best solution to balance economy with excellence was to purchase contemporary works of art directly from the studios of artists, including those who lived and worked in the region.[143] He did not necessarily see this approach as long term, however: 'Such works can be at any time resold or exchanged, and replaced by those inscribed with loftier names when we can afford it.'[144] It is worth noting that if works of art had been bought and sold in this way, Leeds Art Gallery would not have benefited from some of its earlier purchases becoming those 'loftier names' in time, such as two works by the now popular Leeds-born painter John Atkinson Grimshaw (1836–93)—*Iris* and *Nightfall down the Thames*—purchased in 1897.

Funding, construction and early acquisitions

The question of how the collection was to be funded remained open until Harding encouraged the committee responsible for the local celebrations of Queen Victoria's Golden Jubilee to consider combining their activities. The committee had originally planned for the money raised through public subscriptions to support the Imperial Institute in London and the Yorkshire College in Leeds. Harding persuaded the committee to add the art gallery to their list of beneficiaries by promising to donate £1,000, to be matched by his father Thomas Richards Harding (1812–95).[145] This was fortunate timing and further evidence of Harding's opportunism and ability to connect different

141 Bodington and Miall, 'The New Municipal Art Gallery in Leeds', p. 8.
142 Bodington and Miall, 'The New Municipal Art Gallery in Leeds', p. 8.
143 John J. Willson, 'The New Municipal Art Gallery in Leeds', *Leeds Mercury*, 10 December 1886, p. 8.
144 Willson, 'The New Municipal Art Gallery in Leeds', p. 8.
145 'The Queen's Jubilee', *Leeds Mercury*, 4 February 1887, p. 3.

interest groups towards a common—and often mutually beneficial—end result. By June of 1887 £4,400 had been raised for Leeds Art Gallery through the Jubilee Fund.[146]

Meanwhile construction had finally begun, with a contract tendered to the joiners and builders Craven and Umpleby of nearby New Briggate.[147] Before work had even started the budget had increased to £8,924, with the extra £945 required for additional roofing over the central court and gallery for the permanent collection. The project was not without complications: in November 1887 three of the workmen were injured by a falling crane that had been lifting a large iron pillar.[148] In July 1888 the joiners went on strike, for reasons that were not recorded in the press, only to be replaced by a new set of workers supplied by the contractor, to ensure that the gallery could open on 3 October 1888.[149]

Plans for the opening gathered pace over the summer of 1888. Harding began to purchase paintings for the permanent collection, beginning with *Scotland Forever* (1881) by Elizabeth Thompson (1846–1933).[150] This was quickly followed by three paintings by the recently deceased Leeds-born artist John William Inchbold (1830–88): *On the Lagoon: Venice, Near St Raphael* and the watercolour *On the Road to Sepey (Switzerland).*[151] In the same year he added Sydney Herbert's *A Dream of Ancient Athens* (1881). Harding's substantial donation and leading role in the establishment of the art gallery appears to have afforded him licence to purchase works of art for the collection according to his own preferences and judgement. His father also fulfilled his promise to contribute by donating *Noah's Sacrifice* by Daniel Maclise (1847), *A Turkish Warship on Fire* (1868) by Constantin Bolonachi and *Cattle in the Meadows* by Thomas Sidney Cooper. John Barran MP (1821–1905) followed by presenting Gustave Doré's *L'Aube, Souvenir des Alpes* (1877) to the collection, then valued at £500.[152] These works were to be joined by the collection of paintings acquired for the town and displayed at Leeds Town Hall over the previous thirty years, which were cleaned for the occasion.[153]

146 'Her Majesty's Jubilee', *The Times*, 27 June 1887, p. 7.

147 'Leeds Fine Art Gallery and Museum', *Leeds Mercury*, 10 March 1887, p. 8.

148 'Accident at the Leeds Fine Art Gallery', *Leeds Mercury*, 30 November 1887, p. 5.

149 'Leeds Fine Art Gallery and the Joiners Strike', *Leeds Mercury*, 25 July 1888, p. 5.

150 'Leeds Fine Art Gallery', *Leeds Times*, 2 June 1888, p. 3.

151 'Leeds Fine Art Gallery', p. 3.

152 'Gifts and Loans to the Leeds Fine Art Gallery', *Leeds Times*, 4 August 1888, p. 3.

153 'Leeds Fine Art Gallery', *Leeds Mercury*, 10 July 1888, p. 5.

While decisions about the building and its operation processed through the Museum and Arts Committee, to the Free Library Committee, to the Town Council, the composition of the permanent collection was at this time largely dictated by the individuals who chose to donate funds or works of art directly to the gallery. The *Leeds Mercury* noted: 'A large number of the contributors reserved themselves the right of selecting the pictures which their money was intended to purchase. This was a very reasonable condition, and could give rise to no possible quibble.'[154] That the ability to exercise personal preference followed the offer of financial support was not yet considered an ethical problem. It appeared to this class of people entirely natural that their socio-economic status bought them choice, which they were habituated to using over their employees, readership and ratepayers to reinforce their individual and collective authority. The situation was a continuation of the established approach to making temporary art exhibitions in the town, which relied on existing social, political and economic networks to solicit donations. Donors generally professed that their decisions had been taken in the public interest, although what constituted the public interest was highly mutable and contingent on their own beliefs and priorities.

In addition to petitioning his peers for donations to the permanent collection, Harding was actively engaged in securing loans from individuals and institutions to fill the rest of the gallery. South Kensington pledged six cases and nine screens, alongside 'some interesting examples from the Sheepshanks collection of pictures, but not from the particular group which, according to inaccurate local tradition, was once offered to and refused by the Leeds municipality'.[155] There was to be a strong showing from Royal Academicians, including William Quiller Orchardson's *Napoleon on Board the Bellerophon* (1880), a design for a fresco by Frederic Leighton and works by Charles West Cope and Edward Armitage, both artists with Leeds connections.[156] Requests for loans were also made to the corporations of local rivals Manchester, Liverpool and Birmingham. In a significant coup for the gallery, Queen Victoria agreed to lend the popular representation of the Crimean War, *The Roll Call* by Elizabeth Butler, first exhibited at the Royal Academy in 1874.[157]

In addition to the development of the building and the collection, plans were being made for the opening ceremony. The committee had

154 'The Leeds Fine Art Gallery', *Leeds Mercury*, 15 September 1888, p. 12.

155 'Gifts and Loans to the Leeds Fine Art Gallery', p. 3.

156 'Gifts and Loans to the Leeds Fine Art Gallery', p. 3.

157 'The Queen Lends a Picture to the Leeds Art Gallery', *Leeds Times*, 1 September 1888, p. 8.

wanted Sir Frederic Leighton (1830–96), then president of the Royal Academy of Arts, for this occasion, but he was due to be in Europe during the autumn.[158] The present mayor, Archibald 'Archie' Witham Scarr (1827–1904), agreed to host a *conversazione* in the evening and invitations were sent to local dignitaries and the artists Lawrence Alma-Tadema (1836–1912), William Powell Frith (1819–1909) and Hubert von Herkomer (1849–1914).[159] Arrangements were also being made for the public opening. During the first month tickets were to cost 1s. from 10 a.m. to 5 p.m. and 6d. from 5 p.m. to 9 p.m., after which the prices were to be reduced to 6d. in the daytime and 3d. in the evening. A season ticket was available for 5s., valid until the end of the inaugural exhibition on 29 December 1888.[160]

By September 1888 the building was nearing completion and the local periodical press began to preview the new art gallery, 'with a measure of curiosity, not unmingled with some slight remaining doubts'.[161] These remaining doubts coalesced around the associated challenges of the site and the limited resources allocated to the scheme. The *Leeds Mercury* noted:

> That the exterior is not at all striking, the architect himself would be the first to admit, but it must be remembered that throughout the work the committee and the architect alike have been haunted by the grim shade of "economy," and that when the scheme was first proposed it was advanced as a point in favour of its adoption that the fact of the building having to be erected in a bye-street would obviate all necessity for external ornament or ambitious design.[162]

The plain exterior of Bolton Wood stone and pressed brick was no surprise, but the interior appeared to exceed most expectations, particularly in the provision of natural light.

Plaster casts of antique statuary were being installed in the former reading room and the space was judged to form 'a splendid vestibule to the main building', which perhaps suggested that it was not an entirely successful integration (Figure 7.6).[163] The visitor then passed into the core

158 'The Leeds Fine Art Gallery', *Leeds Mercury*, 15 June 1888, p. 8.
159 'Leeds Fine Art Gallery', *Leeds Mercury*, 10 July 1888, p. 5.
160 'Leeds Fine Art Gallery', *Leeds Mercury*, 8 September 1888, p. 7.
161 'The Leeds Fine Art Gallery', *Leeds Mercury*, 15 September 1888, p. 12.
162 'The Leeds Fine Art Gallery', p. 12.
163 'The Leeds Fine Art Gallery', p. 12.

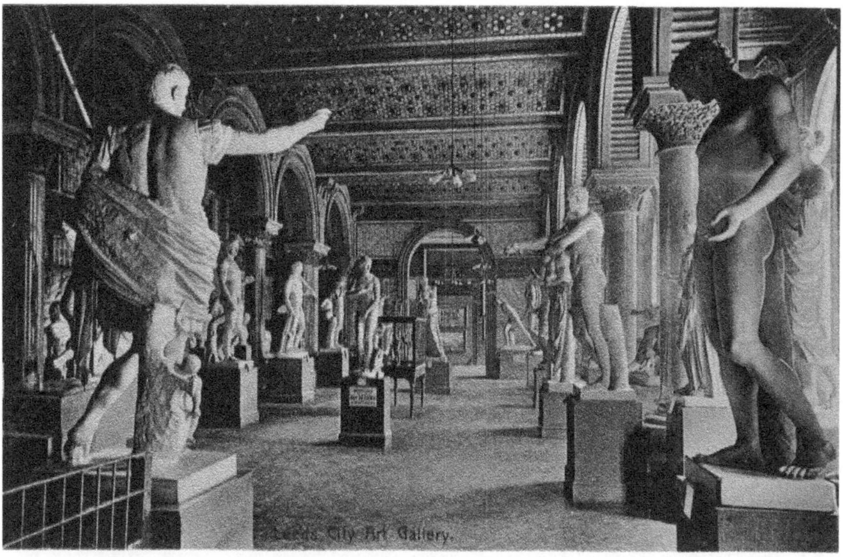

Figure 7.6 William Ritchie and Sons Limited, *Leeds City Art Gallery*, c. 1916, photographic postcard. Collection of the author

vestibule of the new building, which featured a purpose-built niche to accommodate a marble statue of Queen Anne by Andries Carpentière (Anglicised to Andrew Carpenter) (c. 1677–1737). The sculpture was originally commissioned for the exterior of the Moot Hall on Briggate in the centre of Leeds, at the expense of the merchant Alderman William Milner, and was unveiled in 1713. It was restored by Leeds Corporation and transferred to the Old Corn Exchange in 1828 and then transferred to Leeds Town Hall in 1868. In 1887 it was restored once more at the expense of Sir Frederick Milner and other members of his family and relocated to Leeds Art Gallery to mark the Jubilee of Queen Victoria.

From the vestibule the visitor moved into the staircase hall and turned right into the Central Court at the heart of the new building (Figure 7.7). A double-height space with balconies at the first floor, it was decorated with a fountain from the nearby Burmantofts Faience Works and evergreens. It was surrounded to the north and east by two arcades intended to house museum collections connected with the industries of Leeds, alongside temporary displays of loaned objects from South Kensington, which included ancient and modern ironwork, textiles, pottery and domestic utensils from China, Japan, India, Iran,

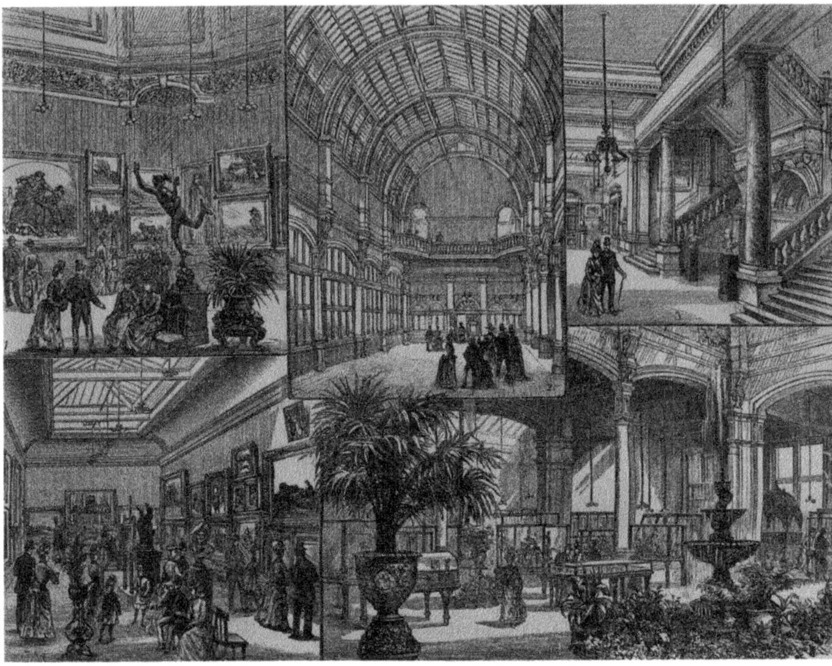

Figure 7.7 F.W., *Municipal Art Gallery and Museum, Leeds*, from *The Illustrated London News*, 27 October 1888, wood engraving. Collection of the author

Italy and Germany.[164] Returning to the vestibule, the visitor turned right into the Queen's Room, intended for the permanent collection (Figure 7.8). The *Leeds Mercury* described the space as 'the handsomest room of the series' as a result of its octagonal roof lantern, panelled ceiling and carved frieze by John Wormald Appleyard (1831–94), which incorporated the names of Hogarth, Reynolds, Gainsborough and Turner.[165] All the walls designed for pictures were boarded over and painted maroon to enable the frames to be screwed directly to the walls, rather than hung from picture rails. The decision appears to have been fairly novel and the reason for the design was not elaborated, although we might assume that it was consistent with the economy of the whole building.

164 'Leeds Fine Art Gallery: The Central Court.—VII', *Leeds Mercury*, 17 November 1888, p. 12.

165 'The Leeds Fine Art Gallery', *Leeds Mercury*, 15 September 1888, p. 12.

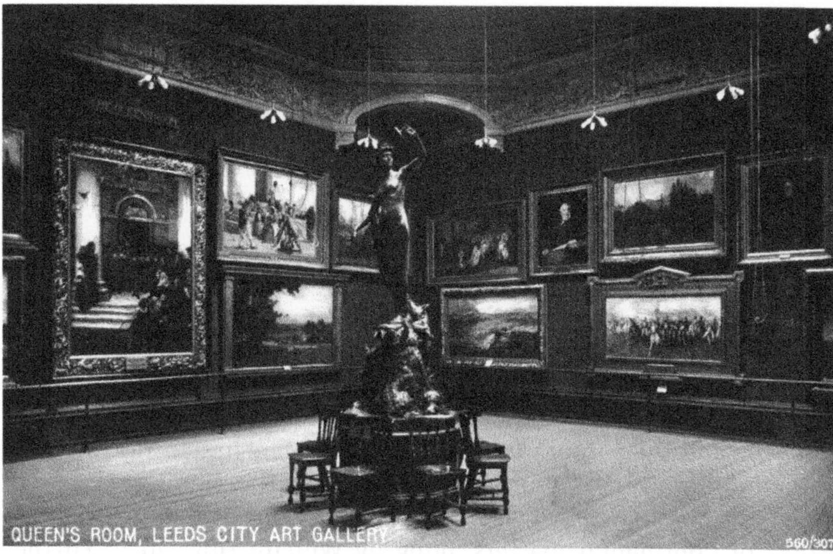

Figure 7.8 William Ritchie and Sons Limited, *Queen's Room, Leeds City Art Gallery*, c. 1915, photographic postcard. Collection of the author

Leaving the Queen's Room, the visitor ascended to the first floor and passed two Doric columns of polished Hopton Wood limestone, which was also used for the balustrades and decorative vases. The first floor contained four top-lit galleries hung with French, Italian and British art from the eighteenth century to the present.[166] In addition to having allowed as much natural light as possible to illuminate the galleries, an experimental system of electric lighting was installed, which included the use of around 1,000 Edison-Swann incandescent lamps of 500 candle power, generated by four Edison-Hopkinson dynamos in the basement.[167]

A few days ahead of the official opening, the *Leeds Mercury* returned to the art gallery for a final inspection before it became a public space. It recognised how contentious the plans had been and how unprepossessing the design:

> Many know how much the site has been made a subject of criticism. If anything proves that the scheme was in good hands, however, the

[166] 'The Leeds Fine Art Gallery', p. 12.
[167] 'The Leeds Fine Art Gallery', p. 12.

> beauty and suitability of the building in its interior arrangements without doubt declare it ... once inside, the visitor perceives not the faintest suggestion of the unsuitability of the gallery to its purpose. For the display of the various works of art, the gallery is certainly equal to that in Trafalgar-square, where the art treasures of the nation are housed.[168]

As the National Gallery in London had cost over £81,000 to construct over fifty years previously, we might question the credibility of this comparison while still recognising the pride in the new institution. The article even had praise for the 'noble apartment' of the former reading room—now the Sculpture Gallery—the relative darkness of which was mitigated by the aforementioned electric lamps.[169]

In the end the corporations of Liverpool, Manchester, Nottingham, Oldham and Wolverhampton lent works of art for the opening exhibition, which was dominated by British art, with only one gallery dedicated to 'works of Continental masters' hung by William Howgate.[170] Further details emerged of the pictures—recently restored by Newton Bros., Mr Broadhead, and Mr Hannah, of Headingley—transferred from the Town Hall, which included Benjamin Robert Haydon's *Mary, Queen of Scots When an Infant* (1842), presented to the town by Andrew Fairbairn to celebrate his time as Mayor of Leeds in 1867.[171] The conservation of the paintings was considered successful and it was said that their appearance 'contrasts very strongly with their shocking appearance formerly, they have been even unrecognisable by Aldermen whose eyes must have looked upon them hundreds of times'.[172] Although the paintings were transferred to Leeds Art Gallery from Leeds Town Hall, the sculptures remained in the Town Hall, perhaps through a combination of the practical difficulties posed by moving large marble statuary even a short distance, the status of sculpture as architectural ornament and the plans to display a collection of plaster casts in the Sculpture Gallery.

168 'The Leeds Fine Art Gallery', *Leeds Mercury*, 29 September 1888, p. 12.
169 'The Leeds Fine Art Gallery', p. 12.
170 'The Leeds Fine Art Gallery', p. 12.
171 'The Leeds Fine Art Gallery', p. 12.
172 'The Leeds Fine Art Gallery', p. 12.

The opening ceremony

The opening ceremony for Leeds Art Gallery went ahead as planned on Wednesday 3 October 1888. The *Leeds Mercury* published a rare illustrated article to mark the occasion, which celebrated the realisation of a long-held ambition. It recognised that Leeds had been 'somewhat tardy' in its provision of a municipal art gallery, in contrast with its early adoption of public libraries.[173] It credited Harding with having given 'concrete form' to the 'feeling in favour of a public fine art gallery', and erroneously claimed that its economical design and construction found 'universal support'.[174] The general impression of the new building was that it had exceeded all expectations, particular with regards to the spacious and elegant interior. There was a sense that a great shame had been lifted and the embarrassments of the past had been erased: 'Leeds has gone back to her old tradition, and it is not too much to say of the inaugural collection that it worthy of the town which organised the exhibition in 1868.'[175] The National Exhibition of Works of Art remained a touchstone even twenty years later. The significance of the exhibitions that had been held in the intervening period was obscured by the perception of financial failure, but they kept the visual arts and their social functions as part of the local public discourse and provided both institutions and individuals with experience to draw from in the formation of a permanent collection.

The *Leeds Mercury* closed its opening day coverage on a highly optimistic note:

> If it fulfils its early promise we shall see it branching out in all directions, becoming by degrees a rich storehouse of well-chosen treasures, in which the artist, the workman, and the child can learn true refinement and the noblest exercise of manual skill. Here all of us, whatever our age or occupation, may find the recreation which leaves us stronger and better than it finds us. Our new Gallery is an effective gain to the dignity of our common life. It will help to rekindle that love for our own city.[176]

According to this position, Leeds Art Gallery not only promised to provide differentiated education and rational recreation, improve

[173] 'The Leeds Fine Art Gallery', *Leeds Mercury*, 3 October 1888, p. 8.
[174] 'The Leeds Fine Art Gallery', p. 8.
[175] 'Our New Gallery', *Leeds Mercury*, 3 October 1888, p. 8.
[176] 'Our New Gallery', p. 8.

industry and manufactures, but offer further intangible benefits of personal and collective dignity and civic pride.

All that the *Leeds Mercury* hoped for in the immediate future of the new gallery was a descriptive guide and the promise of further donations from wealthy patrons. Of the allocation of £800 per annum secured from the Free Libraries budget, half was set aside to pay the interest on the sum spent on the building, of which £5,171 from the Jubilee Fund had been designated.[177] The remaining £400 per annum was sufficient for staff—including the first curator George Birkett—and basic maintenance, but not for the development of the collection, which left the gallery reliant on philanthropy and ticket revenue for further purchases, until the establishment of the Leeds Art Collections Fund in 1912.[178] In addition to the contributions of paintings from local people, the opening ceremony induced at least one notable donation from an artist. Of all the artists from London who had been invited to the opening ceremony, Hubert von Herkomer was the only one to make the journey north. He was quoted as having said of this unfortunate circumstance: 'I deplore the fact that I alone represent my London comrades.'[179] Herkomer encouraged the committee to request donations directly from artists and expressed his support for the new gallery by sending a message by telegraph for a painting of his own to be sent to Leeds immediately. The picture—a small watercolour entitled *The Road Mender receiving the News of the Village*—arrived in time for the mayor's *conversazione* during the evening of the same day.[180] Further support from the capital came from Sir Philip Cunliffe-Owen, who had opened the first exhibition of the Yorkshire Fine Art Society eight years previously. Cunliffe-Owen also urged those present to 'make their gifts while they could participate in the gratification they afforded others' and repeated the misconception that had Leeds been in possession of an art gallery thirty years before, 'one of the finest collections of modern pictures extant in this country would have been presented to it by the late munificent donor, Mr. John Sheepshanks'.[181] That this speculative notion was also held to be true by the director of the South Kensington Museum—the recipient of the Sheepshanks collection—further demonstrates its persistence in both local and national narratives.

177 'The Leeds Fine Art Gallery', *Illustrated London News*, 27 October 1888, p. 11.

178 See Benedict Read and James Lomax, *The Leeds Art Fund: A Centenary History and Catalogue* (Leeds: Leeds Art Fund, 2012).

179 'The Leeds Fine Art Gallery', *Leeds Mercury*, 4 October 1888, p. 8.

180 'The Leeds Fine Art Gallery', p. 8.

181 'The Leeds Fine Art Gallery', p. 8.

Local support beyond Leeds appeared to be strong, with the opening ceremony attended by fourteen Yorkshire mayors.[182] Long accustomed to looking to its northern neighbours with envy, Leeds Art Gallery became an economical model for comparable places to emulate. In nearby York the local newspaper reported that Leeds had 'given to other towns and cities in the north of England an example worthy of imitation'.[183] York had followed a similar cultural trajectory to Leeds, with its own momentum towards a permanent art gallery generated in the last quarter of the nineteenth century through temporary exhibitions of art and industry. It could be argued that the example set by Leeds informed the purchase of the Yorkshire Fine Art and Industrial Institution by York City Council and the subsequent formation of York Art Gallery in 1892.

The public reception

The reception of Leeds Art Gallery by local working-class visitors is more difficult to recover from the available primary evidence. Aimed at the middle classes, the *Leeds Times* reported that 'its erection has been watched by a large number of our working classes with an amount of excitement and eager expectation which has been not the less strongly marked because it has had few opportunities of public expression'.[184] The evidence for their engagement was provided by the contributions made by 'our working men and women' to the art gallery through subscriptions to the Jubilee Fund.[185] During the first day that the gallery was open to the public on 4 October 1888, 125 people paid for admission before 4 p.m. and 185 after 4 p.m. When we recall that ticket prices for the first month of opening cost 1s. from 10 a.m. to 5 p.m. and halved to 6d. from 5 p.m. to 9 p.m., we might reasonably conclude that working people were among the majority who visited in the evening at the cheaper rate. Attendance quickly increased over the course of the opening exhibition, in part stimulated by the planned reduction in the price of tickets to 6d. from 10 a.m. to 4 p.m. and 3d. from 4 p.m. to 9 p.m., and 3d. on Saturdays for the whole day with its extended hours of 9 a.m. to 9.30 p.m. Schools and friendly societies qualified for further discounts of between 10% and 50% depending on the size

182 'The Leeds Fine Art Gallery', *Illustrated London News*, 27 October 1888, p. 11.
183 'Opening of a Fine Art Gallery in Leeds', *York Herald*, 4 October 1888, p. 3.
184 'The Leeds Municipal Art Gallery', *Leeds Times*, 6 October 1888, p. 4.
185 'The Leeds Municipal Art Gallery', p. 4.

of the group.[186] The question of Sunday opening was soon mooted and just as quickly dismissed as premature, despite it having been pointed out that eighteen towns had adopted Sunday opening for their public museums and libraries. In this instance the terms of the debate were less grounded in religious observance and more a concern that free entry would discourage paid visits during the rest of the week and therefore diminish the funds available to develop the permanent collection.[187]

By the close of the first exhibition on 31 December 1888, 75,009 people had visited Leeds Art Gallery, compared for reference to the population of Leeds parish recorded by the 1891 Census as 177,523.[188] Of course, not all the visitors were from the immediate vicinity. The *Leeds Mercury* reported that the Christmas period had enabled visitors from further afield to travel to visit the gallery, which had been particularly busy during the holiday season. It further noted that 'The number of working people who have inspected it since the opening has been very considerable and it is gratifying to report that in no case has any complaint been made of damage.'[189] As has been evident from accounts over the course of the previous half-century, the presence of working-class people at public exhibitions had been both a stated aim and a source of anxiety. Underneath what might have appeared on first reading to have been a congratulatory—if patronising—note was the fear of dirt and perceived poor standards of behaviour at one end of the scale, and active vandalism and crime at the other.

It was estimated that all these visits to the opening exhibition had generated approximately £1,600, of which £500–600 remained after expenses. The *Leeds Mercury* was not impressed by this amount, given the large urban population within a 15-mile radius of Leeds, but conceded that it was more than would have been raised had admission been one penny, or even free of charge. Perhaps conscious of previous instances of financial mismanagement, it urged the committee to be prudent and not to be hasty in spending the surplus to purchase new works of art for the permanent collection.[190] The second exhibition was to be dedicated to artists, with all works for sale. The collapse of the Yorkshire Fine Art Society appeared to be fresh in the mind of the writer, who issued a further note of caution: 'If the Corporation discover

[186] 'Leeds Fine Art Gallery', *Leeds Mercury*, 6 November 1888, p. 5.

[187] 'Sunday Opening Negatived in Leeds', *Pall Mall Gazette*, 1 November 1888, p. 9.

[188] 'Leeds Municipal Art Gallery', *Leeds Mercury*, 31 December 1888, p. 5.

[189] 'Leeds Municipal Art Gallery', p. 5.

[190] Politics and Society, *Leeds Mercury*, 1 January 1889, p. 5.

anything there worth buying, well and good. If not, let them go on funding the money. But what if they have a deficit?'[191] Even after the successful construction of a municipal art gallery for the lowest possible sum, the fear of failure remained.

Conclusion

Being largely concerned with the history of temporary public exhibitions, this is not the place to continue the story of Leeds Art Gallery, but rather to offer some concluding remarks. Giles Waterfield characterised the gallery as one of a 'trickle of nondescript buildings erected in no particular style'.[192] He noted that during this period, art galleries frequently shared spaces and functions, most often with public libraries afforded more space and greater civic prominence.[193] Leeds was certainly part of the 'tradition of understatement' identified by Waterfield in relation to the placement of regional art galleries.[194] Waterfield compared Leeds unfavourably with Liverpool and Glasgow in relation to the comparatively small amount of money spent on its art gallery, and called the town 'notably mean in its treatment of the visual arts... Leeds epitomised municipal frugality' and was 'a national leader in miserliness'.[195]

At the time Thomas Harding located this reluctance in both his peers and the wider public:

> It was said that they were putting up a building in a back street where there would be no light, and that it would be very much better if they would wait—if need be, twenty years longer—to carry out a greater, more ambitious, and therefore more costly scheme. The committee, rightly or wrongly, had gone on the principle that it would be a bad policy to wait an indefinite time until either some generous citizen was moved to do for them what had been done in Liverpool by Sir Andrew Walker, and in Sheffield by the Brothers Mappin, or to wait, also an indefinite time, until public

[191] Politics and Society, p. 5.

[192] Giles Waterfield, *The People's Galleries* (New Haven, CT: Yale University Press, 2016), p. 154.

[193] Waterfield, *The People's Galleries*, p. 166.

[194] Waterfield, *The People's Galleries*, p. 167.

[195] Waterfield, *The People's Galleries*, pp. 253–54.

> opinion had so far grown as to permit of the carrying out of an ambitious scheme out of the public funds.[196]

It is not clear why Harding himself, or at least a consortium of his contacts, did not come forward as these 'generous citizens' to have their names immortalised in the form of an art gallery for Leeds. Only a few years after the opening of Leeds Art Gallery, Harding financed a major sculptural scheme to mark Leeds having been granted city status in 1893.[197] Instigated in 1896, the resulting City Square was unveiled in 1903. The central equestrian sculpture, *The Black Prince* by Thomas Brock (1847–1922), alone cost £5,000, and although Harding convinced fellow councillors to fund two of the bronze statues—James Watt (1898) and John Harrison (1903) by Henry Charles Fehr (1867–1940) were donated by Richard Wainwright and Richard Boston respectively—Harding himself supplied a further eight bronze figures by Alfred Drury (1856–1944) titled *Morn* and *Even*, along with two bronze statues of Walter Farquhar Hook (1902) by Frederick Pomeroy (1856–1924) and Joseph Priestley (1899) by Alfred Drury (1856–1944), together with costly granite plinths and balustrades. It is likely that the timing of this extravagant expenditure was related to the death of Harding's father in 1895 and his subsequent inheritance of land and property in Cambridgeshire, land and cottages adjoining Tower Works, and a share of the residuary estate.[198] Harding's wealth at his death in 1927 was recorded as £365,491 7s., but this does not mean that he had been in a position to donate a large sum of money to the construction and maintenance of an art gallery when he had first proposed the idea over forty years earlier.[199] In an extended article on the history and present state of Leeds published in the *Graphic* in 1885, the author noted that the town 'can boast of fewer merchant princes than are known in connection with other places of equal size, but on the other hand her wealth is widely distributed'.[200] This is not to suggest that Leeds enjoyed any sort of economic equality, but rather that the town had more employers than most and a higher than average proportion of working men who owned their own property.

As this chapter has articulated, the impetus to establish a permanent art gallery and collection for Leeds did not begin with Thomas Walter

196 'The Leeds Fine Art Gallery', *Leeds Mercury*, 4 October 1888, p. 8.

197 Rebecca Wade, 'Benedict Read and the Sculptural Life of Victorian Leeds', *Essays on Sculpture*, 77 (2017), 8–12.

198 'Will of the Late T. R. Harding', *Leeds Mercury*, 8 June 1895, p. 4.

199 Hall, 'Harding, (Thomas) Walter (1843–1927), pin manufacturer and civic art patron'.

200 'Leeds', *Graphic*, 18 July 1885, p. 73.

Harding in 1886. The long history of temporary public exhibitions in Leeds is also a history of attempts to achieve this object. These attempts in the second half of the century were most often appended to another event, institution, building or function, from the first civic collection stimulated by the new Town Hall, to the exhibitions held at the Mechanics' Institute, the founding of the Yorkshire Fine Art Society and the first municipal exhibitions at the Free Public Library brought about by the Society's disintegration. Even the art gallery, once realised, was an architectural addendum to the Municipal Buildings. For the historian R.J. Morris, industrialisation 'brought an inexorable progress from private to public cultural life in Leeds during the last quarter of the nineteenth century'.[201] Leeds Art Gallery was neither a triumph of private philanthropy nor of municipal socialism, but a productive if slightly precarious combination of both. Public support was generated and demonstrated through the Jubilee Fund—as a public subscription dedicated to the building—alongside an ongoing commitment from the Town Council to maintain the operation through a minimal allocation from the rates paid under the Libraries and Museums Act. Leeds was, as the *Graphic* put it in 1885, 'a town that has advanced by thrift'.[202] Private support came in much the same form as it had over the course of the previous fifty years of temporary exhibitions: through the loaning of works of art from private collections for public display. The important distinction was that there now existed a legitimate repository for a permanent collection, which certainly encouraged a steady stream of donations from wealthy individuals, if not the deluge predicted by the promoters of the gallery in its preparatory stages.[203]

The opening of Leeds Art Gallery changed the exhibitionary culture of the town. The voluntary societies that had been so active in previous decades no longer staged ambitious temporary public exhibitions, but their influence continued in a number of ways. Although fine art dominated the discussion, in its early iterations Leeds Art Gallery was also a museum of manufactures, not quite so polytechnic as the

201 R.J. Morris, 'Middle-Class Culture, 1700–1914', in *A History of Modern Leeds*, ed. by Derek Fraser (Manchester: Manchester University Press, 1980), pp. 200–22 (p. 219).

202 'Leeds', p. 73.

203 The first truly transformational single bequest would not occur until 1925, when the Leeds industrialist Sam Wilson left his substantial collection of painting and sculpture, which necessitated the first extension to Leeds Art Gallery. See Leeds City Art Galleries, *Opening of the Art Gallery Extension and of the Sam Wilson Collection by His Excellency The French Ambassador October 12th, 1925* (Leeds: Leeds City Art Galleries, 1925).

exhibitions of the 1830s and 1840s, but still committed to the union of art and industry. It followed too the examples of the Leeds Academy of Arts and the Yorkshire Fine Art Society in its institution of an annual artists' selling exhibition, from which the gallery purchased works of art for the permanent collection with income generated by ticket sales. In these ways Leeds Art Gallery was a culmination of both the successes and failures of the previous generation. The story of exhibitions in Leeds does not, of course, end here but changes in ways that will be explored in the epilogue. Over the course of the following century and more, Leeds Art Gallery would go on to fulfil its promise to 'become a mighty storehouse of art, and a school in which the local genius of the ages to come may receive their first and most abiding impressions'.[204]

[204] 'The Leeds Municipal Art Gallery', *Leeds Times*, 6 October 1888, p. 4.

Epilogue

In the last days of 1889—one year after Leeds Art Gallery had brought its inaugural exhibition to a close—the London newspaper the *Standard* announced that an international exhibition was to be held in Leeds from May to October of the following year. The War Office had given permission for Carlton Grounds to serve as the venue, with Joseph Davies named as the general manager.[1] Applications were solicited to represent engineering, mining, industrial and commercial activities, with prizes to be awarded for models and plans for housing the poor.[2] Local adverts continued into the new year and the international ambitions of the venture quickly became clear when a short article appeared in *Bradstreet's: A Journal of Trade, Finance and Public Economy*, published in New York. The article introduced further sections: electrical and chemical, hygiene, heating and ventilation, water supply, gymnasia, clothing, domestic appliances, food, manufactures and raw materials, education and the fine arts.[3]

The exhibition came together quickly and opened just five months after it had first been announced, on 21 May 1890 (Figure 8.1).[4] Aside from the first appropriation of the term 'international exhibition' in a Leeds context, the exhibition initially appeared to follow broadly the same polytechnic model familiar to local audiences since the late 1830s. But differences soon started to appear. Reporting on the opening of the exhibition, a writer for the *Leeds Mercury* appeared perturbed by the total absence of speeches to mark the occasion, although they

1 'International Exhibition, Leeds', *Standard*, 28 December 1889, p. 1.

2 'International Exhibition, Leeds', p. 1.

3 'Leeds International Exhibition', *Bradstreet's: A Journal of Trade, Finance and Public Economy*, 15 March 1890, p. 169.

4 'Opening of the Leeds International Exhibition', *Leeds Mercury*, 22 May 1890, p. 8.

were somewhat reassured by the programme of music that stood in for them.[5] The article noted that the exhibition was intended to operate as 'a vehicle through which tradesmen may present their wares to public notice, and by which amusements suitable to the summer months may be provided for the people'.[6] This somewhat prosaic intention was a marked contrast to the loftier rhetoric of its predecessors, which tended to promise ennoblement and education, often alongside some form of fundraising for a charitable cause. Davies's International Exhibition was 'of a private description' and organised for profit, although the proceeds generated over the course of three allotted days were donated to the Leeds Workpeople's Hospital Fund.[7]

It is clear that the term 'international exhibition' was used deliberately and specifically to align the exhibition with the lineage of similar ventures staged across the world over the preceding forty years. It was a risky strategy. If it brought additional attention, it also raised expectations. The *Leeds Mercury* reported that the exhibition 'has not assumed the ambitious character associated with the mammoth exhibitions with which we in these modern days have become acquainted'.[8] It did, however, concede that 'by enabling foreigners to take up space it obviates loss which might ensue from lukewarmness nearer our own doors'.[9] Davies had used the same technique the previous year with his 'Great Exhibition' in the nearby city of York in 1889, advertised partially in aid of the County Hospital and the City Dispensary. He was not entirely without connection to the exhibitionary culture of South Kensington, having been involved in the organisation of the Anglo-Danish Exhibition at the Royal Albert Hall in 1888.[10] Despite its title and venue, the exhibition 'had not the official origin of some other similar displays', but secured the patronage of the Prince and Princess of Wales by virtue of the connection with Alexandra of Denmark's home country and its charitable object to rebuild the British Home of Incurables.[11]

When a reporter from the *Leeds Mercury* visited the Leeds International Exhibition for the first time, they noted that 235 firms had submitted goods but not all of them had installed their displays in

5 'Opening of the Leeds International Exhibition', p. 8.
6 'Opening of the Leeds International Exhibition', p. 8.
7 'Opening of the Leeds International Exhibition', p. 8.
8 'Opening of the Leeds International Exhibition', p. 8.
9 'Opening of the Leeds International Exhibition', p. 8.
10 'Opening of the Leeds International Exhibition', p. 8.
11 James Macaulay, ed., *Speeches and Addresses of H.R.H. The Prince of Wales: 1863–1888* (London: John Murray, 1889), pp. 347–48.

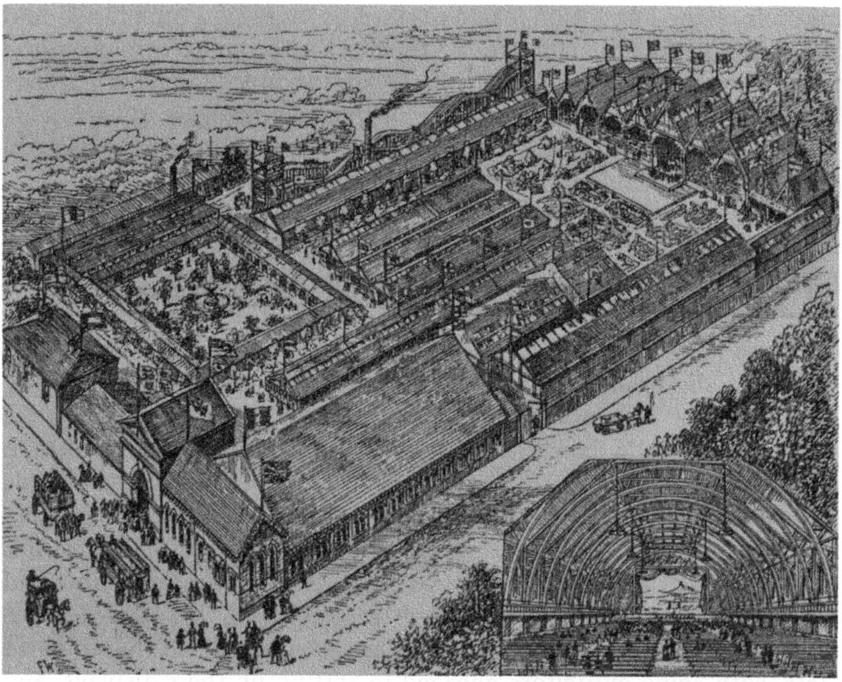

Figure 8.1 F.W., *Bird's-Eye View of the Leeds International Exhibition*, from *The Illustrated London News*, 1890, wood engraving. Collection of the author

time for the opening. The reporter criticised local companies for their lateness but recognised that a delay to the temporary roof structure was partly to blame.[12] The location of the exhibition moved away from the middle-class Mill Hill and the centre of town, towards a residential area between Carlton Hill and Claypit Lane. Home to the drill ground of the Leeds Rifle Volunteers, it was a space associated with military training. The grounds were surrounded by the kind of densely constructed housing associated with the rapid expansion of the town and its working-class population. Although the *Leeds Mercury* described the site as being only 15 minutes' walk from the train station, the exhibition was much closer to the large pool of potential visitors in the back-to-back terraces around Woodhouse and Sheepscar.[13] The

[12] 'Opening of the Leeds International Exhibition', p. 8.
[13] 'Opening of the Leeds International Exhibition', p. 8.

location of the exhibition was described as 'nothing pretentious', and the tightly packed housing around the five acre grounds 'brings out the more forcibly the spaciousness of the site occupied, and the gardens and other decorative adjuncts brought into requisition'.[14] The grounds included a concert hall with a capacity of 1,700, tea gardens, pleasure gardens, a platform for outdoor performances, a switchback railway and a number of other fairground attractions.[15]

When the *Leeds Times* reported on the opening of the exhibition, it could not have been more effusive in its description of the pleasant weather, the 'large and fashionable audience', the refreshments, the excitement of the switchback railway, the variety of the exhibition itself and the promised performances over the coming weeks.[16] The only sense of disappointment came with the absence of the Duke of Newcastle, who had been due to formally open the exhibition and address the crowd.[17] The article concluded: 'Altogether Leeds International Exhibition promised to fulfil its function of combining pleasure and instruction, and there can be little doubt that Yorkshire folks far and near will find their way to the place full many a time and oft.'[18] On the same day that this article was published, the *Leeds Mercury*, by contrast, complained that the exhibition was still not in a finished state, with some exhibitors engaged in setting up their displays, and others yet to start. The article also sounded the first note of caution about the limited number of visitors.[19]

The conflicting interpretations of the success of the exhibition during its first few weeks continued, with the *Leeds Times* countering the *Leeds Mercury* by stating that 7,350 people had attended on the previous Saturday despite poor weather conditions.[20] As the summer season reached its height, so the temporary events associated with the exhibition increased. Towards the end of July it hosted a flower show for three days, which was free for visitors. Held in a separate marquee, prizes worth £70 were offered to exhibitors, although fewer took advantage of the opportunity than had been hoped.[21] More successful

14 'Opening of the Leeds International Exhibition', p. 8.
15 'Opening of the Leeds International Exhibition', p. 8.
16 'Leeds International Exhibition: The Opening Ceremony', *Leeds Times*, 24 May 1890, p. 5.
17 Henry Pelham-Clinton, the 7th Duke of Newcastle-under-Lyne (1864–1928).
18 'Leeds International Exhibition: The Opening Ceremony', p. 5.
19 'The Leeds Exhibition', *Leeds Mercury*, 24 May 1890, p. 7.
20 'Leeds International Exhibition', *Leeds Times*, 19 July 1890, p. 5.
21 'A Flower Show at the Leeds Exhibition', *Leeds Times*, 26 July 1890, p. 5.

were the concerts that took place over the course of the exhibition. Music had been an important feature of temporary polytechnic exhibitions in Leeds since the late 1830s and the Leeds International Exhibition sought to continue the tradition. Although the exhibition operated on a commercial basis, philanthropic activities were not entirely absent. A concert of sacred music was staged in late July to benefit local medical charities, for which an admission charge was made to the grounds. The *Leeds Mercury* reported a large attendance at the event, 'numbering several thousands'.[22] The ticket sales from 15 September and 13 October, both Mondays, were also set to benefit the Leeds Workpeople's Hospital Fund.[23] The first of these three charity days was held on Monday 18 August, with the number of visitors described as 'tolerably good'. It raised £21 19s. 1d. for the Fund, worth approximately £1,800 today.[24]

Early August saw the site return to a more familiar purpose, when almost 6,000 visitors came to see an assault-at-arms performed by men from the Royal Sussex, Loyal North Lancashire, The Border Regiment, Carabiniers, 9th (Queen's) Lancers, 12th (Prince of Wales's) Royal Lancers, 7th (Queen's) Hussars and 15th (King's) Hussars.[25] These live military spectacles were unique to the Leeds International Exhibition, in that they did not have a precedent in earlier polytechnic exhibitions, which featured antiquarian arms and armour but did not show them in use. This form of entertainment appeared to have been particularly popular, with almost 11,000 visitors attending a military mounted competition on the Saturday of the August Bank Holiday weekend.[26]

The *Leeds Times* continued its support into September and early October of 1890. It reported: 'Affairs go swimmingly at Carlton-hill, and some big crowds pass through the gates.'[27] More attention was given to ancillary events than the content of the exhibition, with music, military displays, sports competitions, living statuary and dance continuing to dominate the programme.[28] As the exhibition entered its final month, these attractions continued and the whole event appeared to be on course to be considered a success, with the

22 'Sacred Concert in Aid of the Leeds Medical Charities', *Leeds Mercury*, 29 July 1890, p. 5.

23 'Leeds International Exhibition', *Leeds Times*, 16 August 1890, p. 5.

24 'Leeds International Exhibition', *Leeds Times*, 23 August 1890, p. 5.

25 'Leeds International Exhibition', *Leeds Times*, 9 August 1890, p. 5.

26 'Leeds International Exhibition', *Leeds Times*, 16 August 1890, p. 5.

27 'Leeds International Exhibition', *Leeds Times*, 13 September 1890, p. 5.

28 'Leeds International Exhibition', p. 5.

full list of prize winners published in the *Leeds Mercury* on 25 October.[29] The true situation was soon revealed, however, when the instigator and manager of the Leeds International Exhibition, Joseph Davies, was summoned to the London Bankruptcy Court on 29 October 1890.[30] The case was transferred to the Leeds courts, where it emerged that he had accrued debts of £3,607 and only had assets of £39 17s. 7d.[31] While he claimed that over £10,000 had been generated by the exhibition before expenses, his accounts were missing.[32] Further suspicions about his credibility were raised when it became clear that he had traded under a number of different names; described his profession as an optician; professed to be Jewish when he was not; and erroneously claimed the support of a number of local Leeds men.[33] It was decided that the creditors were unlikely to ever see their debts repaid.[34]

The 1890 exhibition failed from a financial perspective because it was not organised by a local group with a strong social, political and economic network, and, further, Davies did not form an organising committee nor a guarantee fund. Aside from a limited number of days dedicated to raising funds for the Leeds Workpeople's Hospital, the private, commercial nature of the enterprise caused suspicion, which was not relieved by a sense of social purpose—the sale of alcohol at the exhibition, for example, removed it from the moral motive of 'rational recreation' associated with the temperance movement and the 'improvement' of working people. By the last decade of the century the nature of civic culture in industrial urban centres was changing too. Simon Gunn has argued that the form of 'municipal socialism' that thrived in the 1870s and 1880s was no longer dominant and its civic rituals were 'forced to compete for public attention with the attractions of an increasingly commercialised popular culture'.[35] The Leeds International Exhibition of 1890 did not appeal to local civic pride in the same way as its predecessors, nor did it engage with the

29 'Leeds International Exhibition', *Leeds Mercury*, 25 October 1890, p. 9.

30 'Thursday, October 30th, 1890', *Leeds Mercury*, 30 October 1890, p. 4.

31 'The Leeds Exhibition: The Manager's Failure', *Yorkshire Evening Post*, 2 December 1890, p. 4.

32 'The Leeds Exhibition: Meeting of Creditors', *Sheffield Telegraph*, 2 December 1890, p. 2.

33 'How Joseph Davis Ran the Leeds International Exhibition', *Leeds Times*, 31 January 1891, p. 5.

34 'Lively for the Creditors of Joseph Davis', *Leeds Times*, 28 February 1891, p. 5.

35 Simon Gunn, 'Ritual and Civic Culture in the English Industrial City, c. 1835–1914', in *Urban Governance: Britain and Beyond since 1750*, ed. by Robert J. Morris and Richard H. Trainor (Aldershot: Ashgate, 2000), p. 236.

civic structures that had been crucial to the Yorkshire Exhibition of Arts and Manufactures in 1875. As the first significant temporary exhibition organised after the opening of Leeds Art Gallery in 1888, it is possible that the exhibition struggled because it did not articulate a carefully drawn and consistently reinforced relationship between the fine arts and the objects of industry. Once Leeds had a permanent space for the exhibition of fine and decorative arts, the era of the temporary polytechnic public exhibition was over.

The trajectory of public exhibitions in Victorian Leeds began with the voluntary societies that had emerged over the previous two decades. Responding to rapid industrialisation, political reform and demographic change, these subscriber democracies began to define and consolidate middle-class identities. Ideologically opposed to state interference, they advocated the philosophy and practice of self-governance. Polytechnic exhibitions were an arena where the Leeds Philosophical and Literary Society and the Leeds Mechanics' Institution could enact cooperation between different social classes, religious sects and political affiliations, materialise their autodidactic approach to education, and raise funds to secure their position and extend their activities. The establishment of the Leeds School of Design signalled a more pragmatic relationship between state government and local interests. Its committee, derived from the Leeds Mechanics' Institution and Literary Society, were eager to take advantage of the government grants and loans of objects for teaching and temporary public display, but less willing to accept the attendant centralisation and prescriptive curriculum that came with them. This began a longstanding relationship with what would become the Department of Science and Art and the South Kensington Museum, whose collections were regularly circulated to Leeds as part of its design reform agenda.

Temporary public exhibitions were also a very visible means by which the town could define and maintain its cultural and economic position in the region, the country and the wider British Empire. Civic rivalry—most consistently with Manchester, Birmingham and Liverpool—was used to motive Leeds's population to support its exhibitions, to generate civic pride and to construct a sense of local identity through the collective achievements of its artists, industrialists, collectors and institutions. The organisers of these exhibitions, both as individuals and as groups, began as self-appointed authorities who attempted to modify the beliefs and behaviours of visitors, particularly to circumscribe the leisure activities of the working classes. Towards the end of the period, however, elected authorities increasingly took responsibility for the cultural life of the town, which culminated in the establishment of Leeds Art Gallery as a municipal concern. These developments built the structures that continue to support the culture of city, with what is now the largest museum service run by a local authority in the country.[36]

[36] At the time of writing Leeds Museums and Galleries is operated by Leeds City Council and includes nine sites: Abbey House Museum, Kirkstall Abbey, Leeds Art Gallery, Leeds City Museum, Leeds Discovery Centre, Leeds Industrial Museum, Lotherton Hall, Temple Newsam and Thwaite Mills.

Bibliography

Primary Sources

Baines, Edward, *Report Presented to the First General Meeting of the Leeds Mechanics' Institution and Literary Society* (Leeds: printed by Edward Baines and Sons, 1842)

Baines, Edward, *An Alarm to the Nation on the Unjust, Unconstitutional and Dangerous Measure of State Education Proposed by the Government* (London: Ward & Co. and Simpkin, Marshall & Co., 1847)

Baines, Edward, 'On the Woollen Manufacture of England, with Special Reference to the Leeds Clothing District', *Journal of the Statistical Society of London* (1859), 1–34

Baines, Edward, *National Exhibition of Works of Art, Leeds: Baines's Handbook to the Picture Galleries; a reprint, with additions, of critical notices published in the Leeds Mercury* (Leeds: Edward Baines and Sons, 1868)

Baker, Robert, *On the State and Condition of the Town of Leeds in the West Riding of the County of York* (London: Her Majesty's Stationery Office, 1842)

Barber, Fairless, ed., *A Series of Photographs of Objects of Archaeological Interest from the National Exhibition of Works of Art at Leeds, 1868: Taken under the Direction of the Council of the Huddersfield Archæological and Topographical Association with Introductory Remarks and Descriptions by William Chaffers* (London: printed by Whittingham and Wilkins, 1869)

Barnum, P.T., *Struggles and Triumphs: Or, Forty Years' Recollections of P.T. Barnum* (London: Samson Low, Son & Co., 1855)

Birkett, George, *Catalogue of the Permanent Collection of Paintings and Drawings* (Leeds: Leeds City Art Gallery, 1928)

Clarke, Henry G., *A Hand-Book Guide to the Cartoons now exhibiting in Westminster Hall, Designed in pursuance of Notices issued by Her Majesty's Commissioners on the Fine Arts, in furtherance of the object of their enquiry, whether Fresco Painting can be applied with advantage to the decoration of the Houses of Parliament* (London: H.G. Clarke & Co., 1843)

Clarke, Henry G., *A Hand-Book Guide to the Cartoons, Frescos, and Sculpture, Designed in pursuance of Notices issued by Her Majesty's Commissioners on the Fine Arts, in furtherance of their object for ornamenting the New Palace at Westminster. And now exhibiting in Westminster Hall* (London: H.G. Clarke & Co., 1844)

Clarke, Henry G., et al., *A Critical Examination of the Cartoons, Frescos, and Sculpture, exhibited in Westminster Hall* (London: H.G. Clarke & Co., 1844)

Cole, Henry, *The Functions of the Science and Art Department* (London: Chapman and Hall, 1857)

Council of the School of Design, *Third Report of the Council of the School of Design, for the year 1843–44* (London: printed by William Clowes and Sons for Her Majesty's Stationery Office, 1844)

Council of the School of Design, *Fourth Report of the Council of the School of Design, for the year 1844–45* (London: printed by William Clowes and Sons for Her Majesty's Stationery Office, 1845)

Council of the School of Design, *Fifth Report of the Council of the School of Design, for the year 1845–46* (London: printed by William Clowes and Sons for Her Majesty's Stationery Office, 1846)

Curtis, J. Sidney, *The Story of the Marsden Mayoralty; with Sketch of the Mayor's Life. From Authentic Sources* (Leeds: T. Barmby, 1875)

Department of Science and Art, *First Report of the Department of Science and Art* (London: printed by George E. Eyre and William Spottiswoode for Her Majesty's Stationery Office, 1854)

Department of Science and Art, *Second Report of the Department of Science and Art* (London: printed by George E. Eyre and William Spottiswoode for Her Majesty's Stationery Office, 1855)

Department of Science and Art, *Third Report of the Department of Science and Art* (London: printed by George E. Eyre and William Spottiswoode for Her Majesty's Stationery Office, 1856)

Department of Science and Art, *Eleventh Report of the Science and Art Department of the Committee of Council on Education* (London: printed by George E. Eyre and William Spottiswoode for Her Majesty's Stationery Office, 1864)

Ewart, William, *Report from Select Committee on Arts and Manufactures: Together with the Minutes of Evidence, and Appendix* (London: House of Commons Papers, 1835)

Ewart, William, *Report from the Select Committee on Arts and their Connexion with Manufactures; with the Minutes of Evidence, Appendix and Index* (London: House of Commons Papers, 1836)

Exhibition of the Works of Industry of All Nations 1851: Reports by the Juries on the Subjects in the Thirty Classes Into which the Exhibition was Divided (London: William Clowes and Sons, 1852)

The General Infirmary at Leeds: Its Work and Progress, 1767–1911 (Leeds: n.p., 1912?)

Hailstone, Edward, *Portraits of Yorkshire Worthies: Selected from the National Exhibition of Works of Art at Leeds, 1868: With Biographical Notices*, 2 vols. (London: Cundall and Fleming, 1869)

Hailstone, Edward, ed., *Catalogue of the Exhibition of Paintings, Curiosities, Models, Apparatus and Specimens of Nature and Art, at the Music-Hall, Leeds, for the benefit of the Mechanics' Institution; with the names of the contributors* (Leeds: Edward Baines and Sons, 1839)

Hailstone, Edward, ed., *Catalogue of the Second Public Exhibition, of Paintings, Sculpture, Curiosities, Models, Apparatus, Machines at Work, and Specimens of Nature and Art: At the Music-Hall, Leeds, for the benefit of the Mechanics' Institution & Literary Society, with the names of the contributors* (Leeds: printed by Bolland and Kemplay, 1843)

Hailstone, Edward, ed., *Catalogue of the Leeds Third Polytechnic Exhibition, for Providing Public Walks and Baths, containing a magnificent collection of Sculpture, Cartoons, Paintings, Antiquities, Curiosities, Philosophical Apparatus, Mechanical Models, &c. with the names of the contributors, opened June 9th, 1845, at the Music-Hall* (Leeds: printed by Bolland and Kemplay, 1845)

Hole, James, *"Light, more Light!" On the Present State of Education amongst the Working Classes of Leeds and How it Can Best be Improved* (London: Longman, Green, Longman and Roberts, 1860)

Leeds Academy of Arts, *First Annual Report of the Committee of the Leeds Academy of Arts, with List of Subscribers* (Leeds: printed by J. Buckton, 1853)

Leeds City Art Galleries, *Opening of the Art Gallery Extension and of the Sam Wilson Collection by His Excellency The French Ambassador October 12th, 1925* (Leeds: Leeds City Art Galleries, 1925)

Leeds Mechanics' Institution and Literary Society, *Report of the Committee of the Leeds Mechanics' Institution and Literary Society, presented to the annual meeting on the 18th January, 1844, with a list of the Officers and Committee for 1844, &c. &c.* (Leeds: printed by Edward Baines and Sons, 1844)

Leeds Mechanics' Institution and Literary Society, *Report of the Committee of the Leeds Mechanics' Institution and Literary Society, presented to the annual meeting, on the 22nd January, 1845, with a list of the officers and committee for 1845, and an addition Supplement to the Catalogue, containing the additions to the library from Jan. 1843, to Jan. 1845* (Leeds: printed by Edward Baines and Sons, 1845)

Leeds Mechanics' Institution and Literary Society, *Annual Report of the Committee of the Leeds Mechanics' Institution and Literary Society* (Leeds: printed by John Kershaw and Son, 1856)

Leeds Philosophical and Literary Society, *Forty-First Report of the Council of the Leeds Philosophical and Literary Society, at the close of the session, 1860–61* (Leeds: C. Kemplay, 1861)

Leeds Philosophical and Literary Society, *Fifty-Third Report of the Council of the Leeds Philosophical and Literary Society at the Close of the Session 1872–73* (Leeds: printed by McCorquodale & Co., 1873)

Leeds Philosophical and Literary Society, *Fifty-Sixth Report of the Council of the Leeds Philosophical and Literary Society, at the Close of the Session 1875–76. Read at the Annual Meeting, May 2nd, 1876* (Leeds: Printed at the Office of the Yorkshire Post, 1876)

Mayhall, John, *The Annals and History of Leeds, and Other Places in the County of York, from the Earliest Period to the Present Time* (Leeds: Joseph Johnson, 1860)

National Exhibition of Works of Art at Leeds, 1868; Official Catalogue (Leeds: printed by Edward Baines and Sons, 1868)

Norwich School of Design, *A Catalogue of the Casts, Figures, &c. Furnished by Government, for the use of the Students at the School of Design at Norwich, and also of a variety of Articles of French Manufacture, purchased by Government at the Great National Exhibition at Paris in 1844, which have been deposited at the School of Design, Norwich for One Month* (Norwich: printed by Charles Muskett, 1846)

O'Byrne, William R., *A Naval Biographical Dictionary: Comprising the Life and Services of Every Living Officer in Her Majesty's Navy, from the Rank of Admiral of the Fleet to that of Lieutenant, Inclusive. Compiled from Authentic and Family Documents*, vol. 1 (London: John Murray, 1849)

Parsons, Edward, *The Civil, Ecclesiastical, Literary, Commercial, and Miscellaneous History of Leeds, Bradford, Wakefield, Dewsbury, Otley, and the District within Ten Miles of Leeds* (Leeds: F. Hobson, 1834)

Prospectus of the Yorkshire Exhibition of Arts & Manufactures, at Leeds, 2nd edn (Leeds: Fred. R. Spark & Co., 1875)

Redgrave, Richard, *On the Gift of the Sheepshanks Collection, with a view to the Formation of a National Gallery of British Art* (London: Chapman and Hall, 1857)

Reid, James, *Academy Paintings and their Moral Teaching* (Leeds: Richard Jackson, 1884)

Robinson, J.C., *Catalogue of a Collection of Works of Decorative Art: Being a Selection from the Museum at Marlborough House, Circulated for Exhibition in Provincial Schools of Art*, 5th edn (London: printed by George E. Eyre and William Spottiswoode for Her Majesty's Stationery Office, 1856)

Royal Academy of Arts, *Exhibition of the Royal Academy. MMDCCCXLIV. The Seventy-Sixth* (London: W. Clowes and Sons, 1844)

Science and Art Department of the Committee of Council on Education, *Inventory of the Objects in the Art-Division of the Museum at South Kensington, Arranged According to the Dates of their Acquisition. Vol. I. For the Years 1852 to the End of 1867* (London: printed by George E. Eyre and William Spottiswoode for Her Majesty's Stationery Office, 1868)

Smiles, Samuel, *The Diffusion of Political Knowledge among the Working Classes. An Address Delivered before the Members of the Bradford United Reform Club, February 14, 1842* (Leeds: Buckton, 1842)

Stephens, W.R.W., *The Life and Letters of Walter Farquhar Hook D.D., F.R.S.* (London: Richard Bentley and Son, 1881)

Taylor, Richard Vickerman, *The Biographia Leodiensis; or, Biographical Sketches of the Worthies of Leeds and Neighbourhood, from the Norman Conquest to the Present Time* (London: Simpkin, Marshall & Co.; Leeds: John Hamer, 1865)

Waring, J.B., *Extract from the First Report by J.B. Waring, Esq., General Manager and Chief Commissioner of the "National Exhibition of Works of Art," to be held in the New Infirmary at Leeds, in 1868* (Leeds: the Committee, 1867)

Waring, J.B., *Baines's Handbook to the Picture Galleries* (Leeds: Baines, 1868)

West, William, and Edward Baines Junior et al., *A Description of some of the Principal Paintings, Machinery, Models, Apparatus and other Curiosities at the Leeds Public Exhibition* (Leeds: Edward Baines and Sons, 1839)

Wilson, Charles Heath, *Government School of Design Rules Approved and Adopted by the Council, for Attendance, Conduct, and Studies* (London: printed by W. Clowes and Sons, 1843)

Yorkshire Exhibition of Arts and Manufactures, *Yorkshire Exhibition of Arts and Manufactures, Leeds, 1875, Patron: Her Majesty the Queen. Official Catalogue* (Leeds: Inchbold and Beck, 1875)

Yorkshire Fine Art Society, *The First Spring Exhibition 1880* (Leeds: Charles Goodall, 1880)

Yorkshire Fine Art Society, *Catalogue of the Autumn Exhibition, 1880* (Leeds: Charles Goodall, 1880)

Yorkshire Fine Art Society, *The Third Exhibition of the Yorkshire Fine Art Society, Leeds* (Leeds: Charles Goodall, 1881)

Yorkshire Fine Art Society, *Catalogue of the Fourth Exhibition, 1881* (Leeds: Yorkshire Fine Art Society, 1881)

Secondary Sources

Alberti, Samuel, 'Natural History and the Philosophical Societies of Late Victorian Yorkshire', *Archives of Natural History*, 30:2 (2003), 342–58

Altick, Richard, *The Shows of London* (Cambridge, MA: Harvard University Press, 1978)

Arnold, Dana, *Re-presenting the Metropolis: Architecture, Urban Experience and Social Life in London 1800–1840* (Aldershot: Ashgate, 2000)

Arscott, Caroline, '"Without distinction of party": The Polytechnic Exhibitions in Leeds 1839–45', in *The Culture of Capital: Art, Power and the Nineteenth-Century Middle Class*, ed. by Janet Wolff and John Seed (Manchester: Manchester University Press, 1988), pp. 135–58

Baetjer, Katherine, '"Canaletti Painting": On Turner, Canaletto, and Venice', *Metropolitan Museum Journal*, 42 (2007), 163–72

Barber, Brian John, 'Leeds Corporation, 1835–1905: A History of its Environmental, Social and Administrative Services' (unpublished doctoral thesis, University of Leeds, 1975)

Barber, B.J., 'Aspects of Municipal Government, 1835–1914', in *A History of Modern Leeds*, ed. by Derek Fraser (Manchester: Manchester University Press, 1980), pp. 301–26

Beckett, John, *City Status in the British Isles, 1830–2002* (Aldershot: Ashgate, 2005)

Bennett, Tony, 'The Exhibitionary Complex', in *The Nineteenth-Century Visual Culture Reader*, ed. by Vanessa R. Schwartz and Jeannene M. Przyblyski (London: Routledge, 2004), pp. 117–30

Bonser, Kenneth J., *The Music Hall Albion Street: Details of its History from Various Sources* (Leeds: n.p., 1957)

Brake, Laurel, Aled Jones and Lionel Madden, eds, *Investigating Victorian Journalism* (Basingstoke: Macmillan, 1990)

Brears, Peter, *Of Curiosities and Rare Things: The Story of Leeds City Museums* (Leeds: The Friends of Leeds City Museums, 1989)

Briggs, Asa, *Victorian Cities* (London: Odhams, 1963)

Bukantas, Ann, 'John Miller of Liverpool', in *Pre-Raphaelites: Beauty and Rebellion*, ed. by Christopher Newall and Ann Bukantas (Liverpool: National Museums Liverpool, 2016), pp. 74–78

Colls, Robert, and Richard Rodger, eds, *Cities of Ideas: Civil Society and Urban Governance in Britain 1800–2000* (Aldershot: Ashgate, 2004)

Cunningham, Peter, 'The Formation of the Schools of Design, with Special Reference to Manchester, Birmingham and Leeds' (unpublished doctoral thesis, University of Leeds, 1979)

Daunton, Martin, 'Taxation and Representation in the Victorian City', in *Cities of Ideas: Civil Society and Urban Governance in Britain 1800–2000*, ed. by Robert Colls and Richard Rodger (Aldershot: Ashgate, 2004), pp. 21–45

Fielden, Kenneth, 'Samuel Smiles and Self-Help', *Victorian Studies*, 12:2 (1968), 155–76

Foucault, Michel, *The Order of Things: An Archaeology of the Human Sciences* (London: Routledge, 2001)

Grady, Kevin, 'Commercial, Marketing and Retail Amenities, 1700–1914', in *A History of Modern Leeds*, ed. by Derek Fraser (Manchester: Manchester University Press, 1980), pp. 177–99

Gretton, Thomas, '"Art is cheaper and goes lower in France." The Language of the Parliamentary Select Committee on the Arts and Principles of Design of 1835–1836', in *Art in Bourgeois Society, 1790–1850*, ed. by Andrew Hemingway and William Vaughan (Cambridge: Cambridge University Press, 1998), pp. 84–100

Gunn, Simon, 'The Public Sphere, Modernity and Consumption: New Perspectives on the History of the English Middle Class', in *Gender, Civic Culture and Consumerism: Middle-Class Identity in Britain, 1800–1940*, ed. by Alan Kidd and David Nicholls (Manchester: Manchester University Press, 1999), pp. 12–29

Gunn, Simon, *The Public Culture of the Victorian Middle Class: Ritual and Authority and the English Industrial City 1840–1914* (Manchester: Manchester University Press, 2000)

Gunn, Simon, 'Ritual and Civic Culture in the English Industrial City, c. 1835–1914', in *Urban Governance: Britain and Beyond since 1750*, ed. by Robert J. Morris and Richard H. Trainor (Aldershot: Ashgate, 2000), pp. 226–41

Haddock, George, *Some Early Musical Recollections of G. Haddock* (London: Schott, 1906)

Harrison, J.F.C., 'Chartism in Leeds', in *Chartist Studies*, ed. by Asa Briggs (London: Macmillan, 1959), pp. 65–98

Harrison, J.F.C., *Learning and Living 1790–1960: A Study in the History of the English Adult Education Movement* (London: Routledge, 1961)

Huhtamo, Erkki, 'Penetrating the Peristrephic: An Unwritten Chapter in the History of the Panorama', *Early Popular Visual Culture*, 6:3 (2008), 219–38

Huhtamo, Erikki, 'Mechanisms in the Mist: A Media Archaeological Excavation of the Mechanical Theatre', in *Media Archaeology and Intermedial Performance: Deep Time of the Theatre*, ed. by Nele Wynants (Cham: Palgrave Macmillan, 2019), pp. 23–82

Kidd, Alan, and David Nicholls, eds, *The Making of the British Middle Class? Studies of Regional and Cultural Diversity since the Eighteenth Century* (Stroud: Sutton Publishing, 1998)

Kidd, Alan, and David Nicholls, eds, *Gender, Civic Culture and Consumerism: Middle Class Identity in Britain, 1800–1940* (Manchester: Manchester University Press, 1999)

Kitson Clark, Edwin, *The History of 100 Years of Life of the Leeds Philosophical and Literary Society* (Leeds: Jowett and Sowry, 1924)

Kriegel, Lara, *Grand Designs: Labor, Empire, and the Museum in Victorian Culture* (Durham, NC: Duke University Press, 2007)

Kusamitsu, Toshio, 'Great Exhibitions before 1851', *History Workshop*, 9 (1980), 70–89

The Leeds Institute of Science, Art, and Literature: Historical Sketch, 1824–1900 (Leeds: Goodall and Suddick, 1901)

The Leeds Institute of Science, Art and Literature: An Historical Sketch of One Hundred Years Work, 1824–1923 (Leeds: Jowett and Sowry, 1923)

Lomax, James, 'Buying Antiques in Early Victorian Leeds: The 1843 Exhibition', *Furniture History*, 33 (1997), 275–85

Macdonald, Stuart, *The History and Philosophy of Art Education* (London: University of London Press, 1970)

Macleod, Dianne Sachko, *Art and the Victorian Middle Class: Money and the Making of Cultural Identity* (Cambridge: Cambridge University Press, 1996)

McNeil, Ian, ed., *An Encyclopedia of the History of Technology* (London: Routledge, 2002)

McTominey, Andrew, 'Bad Neighbours? Water Supply and the Civic Rivalry of Leeds and Bradford, c.1850–1887', *International Journal of Regional and Local History*, 12:1 (2017), 24–41

Morgan, C.J., 'Demographic Change, 1771–1911', in *A History of Modern Leeds*, ed. by Derek Fraser (Manchester: Manchester University Press, 1980), pp. 46–71

Morgan, Simon, 'John Deakin Heaton and the "elusive civic pride of the Victorian middle class"', *Urban History*, 45:4 (2018), 595–615

Morris, R.J., 'Leeds and the Crystal Palace: A Provincial–Metropolitan Link Bringing Education to Industrial Society', *Victorian Studies*, 13 (1970), 282–300

Morris, R.J., 'Middle-Class Culture, 1700–1914', in *A History of Modern Leeds*, ed. by Derek Fraser (Manchester: Manchester University Press, 1980), pp. 200–22

Morris, R.J., *Class, Sect and Party: The Making of the British Middle Class, Leeds 1820–1850* (Manchester: Manchester University Press, 1990)

Morris, R.J., 'Clubs, Societies and Associations', in *The Cambridge Social History of Britain 1750–1950*, ed. by F.M.L. Thompson, vol. 3 (Cambridge: Cambridge University Press, 1990), pp. 395–444

Morris, R.J., 'Voluntary Societies and British Urban Elites 1780–1850', in *The Eighteenth-Century Town*, ed. by Peter Borsay (London: Longman, 1990), pp. 338–66

Morris, R.J., 'Governance: Two Centuries of Urban Growth', in *Urban Governance: Britain and Beyond since 1750*, ed. by R.J. Morris and Richard H. Trainor (Aldershot: Ashgate, 2000), pp. 1–14

Morris, R.J., and Richard H. Trainor, eds, *Urban Governance: Britain and Beyond since 1750* (Aldershot: Ashgate, 2000)

Mulcahy, Gerardine M., 'An Eminent Sculptor: William Day Keyworth Jun. of London & Hull', *Journal of the Church Monuments Society*, 16 (2001), 126–27

Plunkett, John, 'Moving Panoramas c. 1800 to 1840: The Spaces of Nineteenth-Century Picture-Going', *Interdisciplinary Studies in the Long Nineteenth Century*, 19 (2013), 1–34

Pykett, Lyn, 'Reading the Periodical Press: Text and Context', in *Investigating Victorian Journalism*, ed. by Laurel Brake, Aled Jones and Lionel Madden (Basingstoke: Macmillan, 1990), pp. 3–18

Read, Benedict, *Victorian Sculpture* (New Haven, CT: Yale University Press, 1982)

Read, Benedict, and James Lomax, *The Leeds Art Fund: A Centenary History and Catalogue* (Leeds: Leeds Art Fund, 2012)

Rifkin, Adrian, 'Success Disavowed: The Schools of Design in Mid-Nineteenth Century Britain. (An Allegory)', *Journal of Design History*, 1:2 (1988), 89–102

Rodger, Richard, and Robert Colls, 'Civil Society and British Cities', in *Cities of Ideas: Civil Society and Urban Governance in Britain 1800–2000*, ed. by Robert Colls and Richard Rodger (Aldershot: Ashgate, 2004), pp. 1–20

Rothenstein, John, *Summer's Lease: Autobiography, 1901–1938* (New York: Holt, Rinehart and Winston, 1965)

Sellers, David, *Hidden Beneath our Feet: The Story of Sewerage in Leeds* (Leeds: Department of Highways and Transportation, 1997)

Strickland-Constable, Miranda, *Early Days at Leeds City Art Gallery* (Bradford and London: Lund Humphries, 1974)

Swallow, Deborah, 'The India Museum and the British-Indian Textile Trade in the Late Nineteenth Century', *Textile History*, 30:1 (1999), 29–45

Taylor, David, *The Remarkable Lushington Family: Reformers, Pre-Raphaelites, and the Bloomsbury Group* (Lanham, MD: Lexington Books, 2020)

Thornton, David, 'Mr Mercury—A Biographical Study of Edward Baines with special reference to his role as Editor, Author and Politician' (unpublished doctoral thesis, University of Leeds, 1999)

Thornton, David, *The Story of Leeds* (Stroud: The History Press, 2013)

Trainor, Richard H., 'The "Decline" of British Urban Governance since 1850: A Reassessment', in *Urban Governance: Britain and Beyond since 1750*, ed. by Robert J. Morris and Richard H. Trainor (Aldershot: Ashgate, 2000), pp. 28–46

Unwin, R.W., 'Leeds Becomes a Transport Centre', in *A History of Modern Leeds*, ed. by Derek Fraser (Manchester: Manchester University Press, 1980), pp. 113–41

Wade, Rebecca, 'Pedagogic Objects: The Formation, Circulation and Exhibition of Teaching Collections for Art and Design Education in Leeds, 1837–1857' (unpublished doctoral thesis, University of Leeds, 2012)

Wade, Rebecca, 'Benedict Read and the Sculptural Life of Victorian Leeds', *Essays on Sculpture*, 77 (2017), 8–12

Wade, Rebecca, *Domenico Brucciani and the Formatori of Nineteenth-Century Britain* (New York: Bloomsbury Academic, 2019)

Wahrman, Dror, *Imagining the Middle Class: The Political Representation of Class in Britain, c. 1780–1840* (Cambridge: Cambridge University Press, 1995)

Wainwright, Clive, 'Principles True and False: Pugin and the Foundation of the Museum of Manufactures', *Burlington Magazine*, 136 (1994), 357–64

Wainwright, Clive, 'The Making of the South Kensington Museum I: The Government Schools of Design and the Founding Collection, 1837–51', *Journal of the History of Collections*, 14 (2002), 3–23

Waterfield, Giles, *The People's Galleries* (New Haven, CT: Yale University Press, 2016)

Wolff, Janet, and John Seed, eds, *The Culture of Capital: Art, Power and the Nineteenth-Century Middle Class* (Manchester: Manchester University Press, 1988)

Woodson-Boulton, Amy, *Transformative Beauty: Art Museums in Industrial Britain* (Palo Alto, CA: Stanford University Press, 2012)

Worrall, David, *Harlequin Empire: Race, Ethnicity and the Drama of the Popular Enlightenment* (London: Routledge, 2015)

Yasuko, Suga, 'Designing the Morality of Consumption: "Chamber of Horrors" at the Museum of Ornamental Art, 1852–53', *Design Issues*, 20 (2004), 43–56

Yorke, Herbert, *A Mayor of the Masses: History and Anecdotes of Archibald Witham Scarr* (Leeds: Morrison and Roberts, 1904)

Index

Adelaide Gallery *see* National Gallery of Practical Science
Albani, Francesco 59
Albert, Prince Consort 81, 92, 155, 172
Albion Street 4
Albion Street Music Hall 13, 17, 19, 22, 23il.1, 35–38, 44, 102, 105, 117, 168, 205
Alexander Street 231
Alma-Tadema, Sir Lawrence 242
Ambler, Thomas 228n99, 231, 235
antiquarianism 45, 48, 49, 50–51, 149, 153, 218, 221, 259
antiquities 27, 28, 44, 50, 76, 86, 103, 114, 149, 160, 182
Appleyard, John Wormald 244
aquaria 169
Armitage, Edward 241
arms and armour 114, 149, 155, 181, 182, 217, 259
artists
  British 36, 88, 146–49, 225, 242
  donations from 248
  groups 117, 126–27, 222
  loans from 22, 24, 43, 102n93, 110, 215
  local 34, 99–100, 124, 188, 210–11, 213–14, 239–41
  non-Western 26–27
  patronage of 216
  sales by 22, 34, 43, 250
  status of 23–24, 53, 55–58, 62–64, 83, 86, 94, 101, 120, 181, 261
  teaching 118
  women 26, 61–62, 105, 240, 241
Athenæum Building 16, 200, 212–13, 218, 237
Austin, John 106–7, 110
autodidacticism *see* Smiles, Samuel

Baines, Sir Edward 7–12, 18–19, 32–33, 35, 41, 65, 76, 178n83, 202, 204–5
Bakewell, William 168, 212, 237
Barran, John 100, 240
Bates, Edmund 222
Baths and Washhouses Act 111
Beck and Inchbold 162–63
Beckett, John 92
Beckett, William 70, 111n127, 130
Beckett and Company 161, 200
Beckett Denison, William 206
Beckett Street Cemetery 184
Behnes, William 92
Bellini, Gentile 147
Berchem, Nicolaes Pieterszoon 187
Birkett, George 248
Birmingham 1, 165, 172, 173
  art union 101
  civic collection 241
  civic rivalry 110, 114, 226, 261
  exhibitions 17, 129
  Museum and Art Gallery 239
  public walks and baths 77
Bischoff, John 191

Boddington Junior, Edwin Henry 187
Bodington, Sir Nathan 236–39
Bolonachi, Constantin 240
Bradford
  Art Gallery and Museum 212, 222, 226
  artists from 100
  exhibitors from 173
  infrastructure 178
  manufacturers 121
  Mechanics' Institution 8
  railway 3, 98
  visitors from 156
British Empire 82, 114, 149, 172, 184, 261
British Institution 53
Brock, Sir Thomas 252
Brodrick, Cuthbert 33n75, 140
Bromley, Alfred 66–67

Calder Marshall, William 154, 155
Caracciolo, Giovanni Battista 193
Caravaggio, Michelangelo Merisi da 22, 191, 193
Carlton Grounds 254
Canal, Giovanni Antonio 60
Carpentière, Andries 243
Carr, John 140
Carracci, Annibale 98
Centenary Street 231, 234
Chaffers, William 149, 152
Chapel Allerton and Potternewton Mechanics' Institute 189
Chartism 33–34, 68
Chemical Society of London 178
China 47, 52, 103–4, 149, 181, 243
cholera 79, 178–79
Chorley, Charles R. 235, 237
Cima da Conegliano, Giovanni Battista 147
City Square 52, 168n44, 252
cloth halls 165–66
  Coloured Cloth Hall 15, 165–68, 170, 172, 174, 180, 194, 196, 211
  Irregulars' Cloth Hall 36–37
  Tom Paine's Hall *see* cloth halls, Irregulars' Cloth Hall
  White Cloth Hall 166
Cole, Sir Henry 78, 131, 214
Commercial Buildings 9n37, 47
concerts *see* music
conservation 58–60, 96, 243, 246
Cooper, Thomas Sidney 80, 240
Cope, Charles West 53, 97, 98, 241
Correggio, Antonio Allegri da 22, 40, 98, 187, 194
Corson, George 237
Craven and Umpleby 249
crime 70, 78, 84–85, 250
Crystal Palace 96, 141, 142, 143, 145
Cundall, Robert 183–84
Cundall and Fleming 148, 152
Cunliffe-Owen, Sir Francis Philip 214–15, 248
Cust, Sir Lionel Henry 148

Davies, Joseph 255–56, 260
Davis, James William 215
Dayson, J.O. 161
Dell, John Henry 187
Department of Science and Art 13, 128–37, 207, 214, 261
design reform 13, 64, 117, 201–2, 207, 218, 236, 261
Dickens, Charles 50
dioramas 47
dissolving views 67, 70, 81, 108
Distin family 109–10
Dixon, A.S. 220
Doré, Paul Gustave Louis Christophe 240
Drury, Alfred 52, 252
Duke of Edinburgh 15, 163, 165, 170, 171i6.1, 172, 175, 196, 215
Durham, Joseph 155

Earle, Thomas 80, 88–90
Eastlake, Sir Charles 125
electricity 19, 29–30, 65, 72–73, 108, 245, 246, 255

Elkington and Company 152
Emsley, William 233

Fairbairn, Sir Andrew 246
Fairbairn, Sir Peter 20n13, 163, 200–5
Fairfaxiana 24, 26, 193, 218
Farnley Hall 22, 24, 44, 193–94, 218
Fawkes, Ayscough 193–94, 218
Fawkes, Francis Hawksworth 22, 24–26, 44, 193
Fehr, Charles Henry 52, 252
Fichel, Benjamin Eugène 187
Fine Arts Commission 14, 88
  see also Houses of Parliament; Westminster Hall
First Opium War 103, 114
Flaxman, John 148, 153
flitch system 167–68
Foley, John Henry 154
fountains 19, 72, 109, 153, 169, 243
Francis, John 92
Fraser and Sugden 237
Frith, William Powell 242

Gainsborough, Thomas 81, 98, 146, 244
Gibson, John 154
Gilbert Scott, Sir George 15, 140, 143
Goldsmiths' Hall 91, 93i3.2
Gormley, Anthony 1
Gott, Benjamin 216
Gott, John 211, 216
Gott, William 216
Great Exhibition of the Works of Industry of All Nations 2, 4, 17, 96, 98n79, 129, 144, 156, 201
Greenwood and Batley 169
Grimshaw, John Atkinson 222, 239
guarantee funds 112, 141, 160, 161, 178, 188, 211, 213, 218, 260
Gullick, Thomas John 215

Hailstone, Edward 148
Halifax 121, 156, 173
Handyside and Company 169
Harding, Thomas Richards 239–40
Harding, Thomas Walter 16, 52, 225–31, 233, 237, 239–41, 247, 251–53
Hargrove, Charles 224
Harrison, John 52, 252
Haydon, Benjamin Robert 80, 100, 246
Heaton, John Deakin 202, 204
Herbert, Sydney 240
Herring Junior, John Frederick 187
Hogarth, William 146, 244
Holdforth, James 38
Hole, James 206–7
Hook, Walter Farquhar 52, 77, 92–97, 252
Houses of Parliament 14, 88, 90, 103, 114
  see also Fine Arts Commission; Westminster Hall
Howgate, William 214, 219, 246
Huddersfield 156, 173, 189, 220
Huddersfield Archæological and Topographical Association 152
Hull 3, 13, 42, 76, 80, 88, 92, 94–96
Hunt and Roskell 153

Inchbold, John William 240
India 14, 15, 26, 27, 47, 52, 128, 149, 155–56, 214, 243
industry 1, 2, 4, 5, 8, 14, 15, 16, 17, 19, 25, 32, 39, 40, 68, 74, 78–79, 91, 94, 97, 117, 124–25, 142, 156, 158, 160, 169, 196, 216, 220, 228, 243, 249, 253–54, 261
  design 122–23
  engineering 3
  improvement 129, 238–39, 247–48
  machinery 20, 30–31, 34, 35, 76, 174–77
  manufactures 134, 225
  textiles 3, 166
  woollen 10

John Fowler and Company 174

Keyworth, William Day 52, 92–97
Kitson, James 97, 129, 140, 155, 178n83
Kitson and Company 67, 68

labour 3, 5, 6, 12, 14, 20, 30, 35, 40, 48, 50, 53, 68, 78, 124, 176, 230
Lady Butler *see* Thompson, Elizabeth Southerden
Landseer, Edwin 55, 80
Lara, Georgina 187
Lawrence, Sir Thomas 25, 101–2
Lee, John 223–24, 234, 238
Leeds Academy of Arts 14, 67, 102, 122–28, 136–37, 157, 158, 218, 254
Leeds and West Yorkshire Architectural Society 235, 237
Leeds Art Gallery
  collection 192
  establishment 1, 4, 13, 16, 139, 158, 195–97, 199–254
Leeds Art Union 100–2
Leeds City Council 1, 143n74, 163, 204–6, 208n21, 228–29, 253, 262
Leeds Corporation *see* Leeds City Council
Leeds General Infirmary 14–15, 130n55, 139–46, 146i5.3, 153, 156–58, 185, 188
Leeds Grand Theatre 211
Leeds Mechanics' Institution 5, 40, 127
  buildings 33, 34n77, 140, 206–7
  education 96, 122, 123n26, 207, 209
  exhibitions 13, 15, 35, 41, 77, 118, 121, 130, 135, 159–61, 185, 189, 261
  fundraising 14, 31–33, 43, 46, 49, 71, 74, 76, 77, 117, 130, 136, 194–96, 261
  governance 9, 129
  membership 6, 8, 18–19, 65, 261
  patronage 7
Leeds Mechanics' Institution and Literary Society *see* Leeds Mechanics' Institution
Leeds Municipal Buildings 16, 212, 219, 221i7.4, 222, 224–25, 227, 230–31, 232i7.5, 233, 236–38, 253
Leeds Museums and Galleries 1, 262n36
Leeds Philosophical and Literary Society 5–9, 13, 27, 40, 95, 178n83, 185, 192–93, 195–96, 204, 233–34, 261
Leeds Photographic Society 222
Leeds Public Baths Company 110
Leeds Rifle Volunteers 257
Leeds School of Art *see* Leeds School of Design
Leeds School of Design
  curriculum 123–24
  establishment 13, 117, 127, 207, 261
  exhibitions 14, 118–22, 128–37, 201, 205–7
  teaching collection 96, 160, 207, 217
Leeds School of Practical Art *see* Leeds School of Design
Leeds Society of Artists 222
Leeds Town Council *see* Leeds City Council
Leeds Town Hall 16, 97, 155, 165, 169, 172, 185, 200, 201i7.1, 202–8, 212, 221–22, 224, 237, 238, 240, 243, 246, 253
Leeds Working Men's Association 68
Leeds Workpeople's Hospital Fund 256, 259, 260
Leeds Zoological and Botanical Gardens 110–13
Leighton, Sir Frederic 241, 242
leisure *see* recreation
Libraries and Museums Act 233, 253
Liverpool 1
  art union 101
  civic collection 241, 246

civic rivalry 42, 76, 82, 110, 114, 136n79, 211, 221, 226, 251, 261
collectors 98
exhibitions 13, 105, 107
industry 3
Walker Art Gallery 225, 230, 238, 251
locomotives 65, 69, 97
London
artists from 23–24, 99, 126, 248
collections loaned from 114, 128, 132, 134, 136, 155
influence of 39, 142–45, 148
polytechnic institutions in 19, 30, 47, 72, 108
railways 3, 170
temporary exhibitions in 4, 21, 82, 88, 90, 106, 158
Lunardi, Vincenzo 166
Lupton, Darnton 79

MacDowell, Patrick 154
Maclise, Daniel 240
Maher and Sons 172
Manchester 1, 100, 156
art union 101
civic collection 241, 426
civic rivalry 42, 76, 82, 110, 114, 140, 226, 261
exhibitions 13, 17, 139, 140–41, 145, 148, 158
Free Library 205
industry 3
public walks and baths 77
railways 3, 98
Mann, Joshua Hargrave Sams 187
Marochetti, Baron Carlo 155
Marsden, Henry Rowland 15, 159, 161, 174–75
Marshall, Arthur 209, 216
Marshall, John 216
Marshall, Victor 216
Martin, John 56, 80, 81, 97, 98, 102, 102n93
Massey-Mainwaring, William Frederick Barton 222
McLellan Galleries 141–42
Miall, Louis Compton
Midwood, William Henry 187
Mill Hill 4, 7, 38, 98, 127, 185–86, 224, 257
Miller, John 98
Milner, Sir Frederick 243
Milner, William 243
Moot Hall 243
municipal socialism 228–29, 253, 260
Murillo, Bartolomé Esteban 102, 191, 192
Museum of Manufactures *see* South Kensington Museum
Museum of Ornamental Art *see* South Kensington Museum
music 31, 33, 36, 37, 65, 69, 71, 72, 109–10, 168, 169, 171, 178, 186, 194, 213, 220, 256, 259

National Gallery 142, 147, 246
National Gallery of Practical Science 19, 30
New Zealand 52, 103–4
Newcastle 17, 73
newspapers 2, 7–8
Niemann, Edmund John 187
Nixon, Samuel 91–92
Noble, Matthew 92, 155, 202, 203i7.2
Nollekens, Joseph 92
Northern Society for the Encouragement of the Fine Arts 34, 36, 53, 127, 134n74, 191
Nursey, Claude Lorraine 123

Orchardson, William Quiller 241
Ordish, Rowland Mason 143–44

Pacific Islands 182–83
Paganini, Niccolò 37
panoramas 70, 106
Park Estate 5
Park Row 4–7, 33, 95, 108, 163, 233, 237

Park Square 4, 78, 237
Peel, Sir Robert 81
Philosophical Hall *see* Leeds Philosophical and Literary Society
photography 15, 57, 107–8, 134, 148, 152, 160, 177, 178, 180, 222
plaster casts 96, 118, 124, 134, 153, 192, 216, 217, 242, 243i7.6, 246
Polytechnic Institution 19, 30, 47, 72, 74n123, 108
population 2, 17, 35, 38, 79, 140, 221, 250
Pre-Raphaelites 98, 146
Priestley, Joseph 52, 252
Provis, Alfred 187
public health *see* sanitation

Queen Victoria
  Golden Jubilee 16, 239, 243
  loans from 24, 132, 149, 152, 181, 241
  patronage 15, 162, 196, 202
  portraits of 92, 144, 153, 155, 203i7.2
  visits to Leeds 164, 172, 196

railways 3, 28, 65, 71n107, 81, 98, 108, 118, 126, 132, 165, 178, 196, 258
Ramsden, Archibald 163
recreation 10, 18, 39, 40, 45, 78, 81, 176, 178, 204, 210, 247–48, 260, 261
Reform Acts 4, 7–8
Reid, James 229–30
Reni, Guido 60, 61, 81, 105n105, 194
restoration *see* conservation
Reynolds, Sir Joshua 146, 194, 244
Rhodes, John 53–55, 100, 187
Rhodes, Joseph 23i1.1, 100
Ribera, Jusepe de 192–93
Richardson, Thomas Miles 187
River Aire 5, 79
Robinson, George 214
Romney, George 148, 194
Rothenstein, Sir John Knewstub Maurice 192–93
Roundhay Park 111, 165
Royal Academy of Arts 25, 53, 59, 67, 80, 88, 92, 94, 117, 123–25, 127, 142, 147, 153–55, 211, 241, 242
Royal Albert Hall 256
Royal Cornwall Polytechnic Society 91
Royal Polytechnic Institution *see* Polytechnic Institution
Rubens, Sir Peter Paul 22, 102, 194
Ruskin, John 8, 64, 237

Salford Royal Museum and Public Library 205
Salmon, John Cuthbert 187
Sandwith, Thomas Backhouse 195
sanitation 14, 78–80, 110, 114, 161–62, 178–80, 196
Savoldo, Giovanni Girolamo 147
Scarr, Archibald Witham 163–64, 242
Scharf, George 148
science 10, 14, 21, 24, 29, 39–40, 42, 48, 49, 52, 68, 71–72, 74, 86, 107, 159, 177–79, 196, 206
Scott, Sir George Gilbert 15, 140, 143
sculpture 14, 15, 80, 81, 85–97, 99–100, 110, 114, 120–21, 123, 135, 141, 142, 145, 149, 153–55, 158, 160, 180, 192, 200–1, 205, 208, 210–11, 216, 218, 221, 227, 231, 233–34, 238, 243, 246
  public sculpture 1, 52, 66–67, 252
self-help *see* Smiles, Samuel
Sèvres 120, 132, 133i4.1, 222
Sheepshanks, John 227–28, 241, 248
Sheffield 17, 122n24, 135, 173, 251
Shirley, Henry 187
Smiles, Samuel 8–9, 11–12, 164
Snyders, Frans 55, 194
South Kensington Museum 13, 15, 136, 142, 149, 152, 181, 201, 204, 207, 214, 222, 224, 227–28, 233, 238, 241, 243, 248, 261

South Parade 9n37, 33, 34n77, 43, 117, 118
Spark, Frederick Robert 228
Stanhope, Philip Henry 148

taste 23, 30, 32, 34, 40, 42, 45, 46, 49, 50, 51, 63–64, 68, 78, 99, 104–5, 113, 124, 126, 129, 142, 182, 184, 202, 206, 216, 217, 225
Tatham, George 218
Taylor, Matthew 170
Thompson, Elizabeth Southerden 240, 241
Thornycroft, Mary 155
Thorp, William Henry 237–38
Throp, John 211, 213
tickets 27, 38–39, 47, 65, 68–71, 76, 83–85, 102, 114, 136, 156, 157i5.8, 162, 167, 171, 193–95, 214, 242, 248–49, 254, 259
transport *see* railways
Trumble and Cooke 119
Turner, Joseph Mallord William 24, 26, 58, 193–94, 244
Tweedale, John 228n99, 231, 235

urban history 2–3, 5, 8–9, 25, 38–39, 68, 78–80, 165–66, 188, 228–29, 260–61

van Dyck, Sir Anthony 22, 53, 56, 57, 194
van Rijn, Rembrandt Harmenszoon 22, 80, 81, 98, 100
Victoria Museum *see* Cundall, Robert
voluntaryism *see* Baines, Sir Edward
von Herkomer, Sir Hubert 242, 248

Wakefield 96, 98, 156
Walker, Sir Andrew Barclay 225
Walker Art Gallery *see* Liverpool
Waller, Richard 100, 124, 125, 214, 218
Walley, Joseph 184–88
Walton's Music Saloon *see* South Parade
Ward, William Sykes 177–78
Waring, John Burley 141–42, 147, 151–52, 153
Watson, John Forbes 155–56
Webster, Thomas 62–64
Wedderburn, Lancelot 188–92, 212
West Riding Geological and Polytechnic Society 178
Westminster Hall 80, 87i3.1, 88–90, 92
see also Fine Arts Commission; Houses of Parliament
William Lupton and Company 176i6.3
Wilkie, David 101–2
Williams, John 182–83
Willson, John Joseph 212–13, 239
Wilson, Charles Heath 118–22
Woodhouse Moor 111, 172
Worsnop, Charles B. 133
Wortley Working Men's Institute 236
Wouvermans, Philips 102
Wright, Joseph 53, 54
Wyatt, Matthew Digby 141

York Art Gallery 249
Yorkshire College 236, 238, 239
Yorkshire Fine Art and Industrial Institution 249
Yorkshire Fine Art Society 16, 200, 210–20, 224, 237, 239, 248, 250, 253, 254
Yorkshire Union of Mechanics' Institutes 8, 206

Printed and bound by CPI Group (UK) Ltd, Croydon, CR0 4YY

13/09/2026

14961098-0006